T0321597

Liquidator
The Chernobyl Story

Other Related Titles from World Scientific

The Soviet Atomic Project: How the Soviet Union Obtained
the Atomic Bomb
by Lee G Pondrom
ISBN: 978-981-3235-55-7

Atom Projects: Events and People
by Boris Ioffe
ISBN: 978-981-3145-93-1
ISBN: 978-981-3145-94-8 (pbk)

Crossing the Red Line: The Nuclear Option
by Gerald E Marsh
ISBN: 978-981-3276-82-6

Basics of Transport and Storage of Radioactive Materials
edited by Toshiari Saegusa, Gilles Sert, Holger Völzke and Frank Wille
ISBN: 978-981-3234-03-1

Liquidator

The Chernobyl Story

Sergei Belyakov

World Scientific

NEW JERSEY · LONDON · SINGAPORE · BEIJING · SHANGHAI · HONG KONG · TAIPEI · CHENNAI · TOKYO

Published by

World Scientific Publishing Co. Pte. Ltd.

5 Toh Tuck Link, Singapore 596224

USA office: 27 Warren Street, Suite 401-402, Hackensack, NJ 07601

UK office: 57 Shelton Street, Covent Garden, London WC2H 9HE

Library of Congress Cataloging-in-Publication Data

Names: Belyakov, Sergei, 1956–   author.

Title: Liquidator : the Chernobyl story / Sergei Belyakov.

Description: Singapore ; Hackensack, NJ : World Scientific Publishing Co. Pte. Ltd., [2018] |
    Includes bibliographical references and index.

Identifiers: LCCN 2018017440| ISBN 9789813227415 (hardcover ; alk. paper) |
    ISBN 9813227419 (hardcover ; alk. paper) | ISBN 9789813228689 (pbk ; alk. paper) |
    ISBN 9813228687 (pbk ; alk. paper)

Subjects: LCSH: Belyakov, Sergei, 1956– | Chernobyl Nuclear Accident, Chornobyl',
    Ukraine, 1986--Personal narratives. | Radioactive waste sites--Cleanup--Ukraine. |
    Nuclear accidents--Health aspects--Ukraine. | Nuclear power plants--Ukraine--Safety measures.

Classification: LCC TK1362.U38 B485 2018 | DDC 363.17/99094777--dc23

LC record available at https://lccn.loc.gov/2018017440

British Library Cataloguing-in-Publication Data

A catalogue record for this book is available from the British Library.

For any available supplementary material, please visit
https://www.worldscientific.com/worldscibooks/10.1142/10636#t=suppl

Desk Editor: Ng Kah Fee

Typeset by Diacritech Technologies Pvt. Ltd.
Chennai - 600106, India

Printed in Singapore

Dedication

To Mom and Dad — I know that you would
have enjoyed reading this book.

To Laura — love, many great things in my life
wouldn't have happened without you.

To thousands of my compatriot liquidators — we
did what we could to save the world.

Special thanks to Mrs. Mary Halpin and Mr. Andrew Rooke
for their invaluable help with the manuscript preparation.

Every event, scene, action, description of the background, and author's personal accounts, perception of the happenings, feelings are depicted as accurately as possible. To protect the identities of actual persons, alternative names are used in the book.

Preface

In the middle of the last century, American industrial psychiatrists Frank and Lilian Gilbreth came up with a set of fundamental motion elements that are typical for every worker to perform a manual operation or task. Their set contains eighteen elements called therbligs, a reversal of the scientists' last name with the last two letters transposed. The therbligs are split into two groups — eight effective and ten ineffective. I actually disagree with some of those assigned by Frank and Lilian as ineffective. I would oppose the assignment of "Position", "Inspect", "Search", "Select", and "Plan" to the group of ineffective elements; and somewhat agree with "Rest" to be classified as inefficient, although the sheer meaning of "Rest" does not fall into a category of motion anyway.

I don't think that knowing about the existence and especially usefulness of therbligs during my stint in Chernobyl would have helped me to do the job better. In the scorching summer of 1986, in the rush, stress and incredible intensity of those days and weeks, there was no place and time for taking a break and analyzing the effectiveness or the lack thereof of my motions, every one of which was devoted to the one and only goal — stop the spread of radiation from the exploded nuclear reactor.

I shared that devotion with many others. We were called the "liquidators". In Fukushima's early days of tragedy, someone came up with another name of a similar job — the "nuclear jumpers". I must admit, it sounds much swankier, although it does not change the scope of what's behind them both.

In the official documents of every liquidator there is a wording " ... participated in the alleviation of consequences of the CNPP accident in so-and-so year". I hail the opacity of such a description, created by ingenuity of the Soviet bureaucracy, but it never sat well with me. Whoever came up with it clearly had no clue of what exactly we went through, we meaning all liquidators, regardless of the involvement, small or large, continuous or incremental, efficient or not so much (therbligs!).

One cannot simply participate in it. One must live through that, be a vital, important, fully engaged small component of the colossal task, which had no analogues in the history of mankind. Atop of that, we were officially called for the Army Reserve duty, with a lot of the military reserve nonsense plainly interfering with our everyday work at the station (I will address this point later in the book).

The importance and the value of those five therbligs in question were not challenged in the summer of 1986; at least, not when I was leading the teams of 10 to 25 troops during the clean-up shifts at a heavily contaminated nuclear plant. I just did not realize that our actions back then could be analyzed by (or could be matched against) the therbligs. All that we were doing was locked dead on in a short expression that I have heard from many liquidators, who were interviewed by the media after the Chernobyl saga was over.

"Someone had to do it. If not me, who else?"

I live now in that part of the world where the whole event and the threat of it for mankind went relatively unnoticed, and the younger friends of mine, my coworkers, my students, feel a bit uncomfortable when somehow the conversation focuses on the Chernobyl accident. Some of them plainly do not know what this name is associated with; some of them were born after the accident, thousands of miles away, so they are oblivious to what happened, more so, could have happened if our clean-up efforts fell short.

I don't talk about my liquidator's past much. However, I am concerned that soon the name "Chernobyl" will be linked in our minds more to a popular computer game, than to a landmark of the real planetary scale disaster.

And this is another reason why I decided to write this book.

Sergei Belyakov

Contents

"Praemonitus praemunitus" *(lat.)*
"Forewarned is forearmed"
(E. D. Hirsch, Jr., *et al.*
"The New Dictionary of Cultural Literacy"
3rd Edition, Boston, 2002)

PART 1

"Are You Tired of Your Life?"

In late April of 1986, two of my friends and I went for a fishing trip to the upper Dnieper river region, *Farther Lakes*, a beautiful natural setting, where the main river stream splits into dozens of minor ones, and often loses the urgency to go south, forming innumerable small lakes and springs filled with clean fresh water. Branches, lakes and springs were a true home to the wildlife, from wolves, foxes and even bears to weasels, ferrets, and countless flocks of cranes, herons, geese, ducks and of course every kind of fresh water fish known in Eastern Europe. Farther Lakes were the place of our yearly pilgrimage at the end of the spring semester — all three of us were teaching at the Institute of Chemical Technology in Dnepropetrovsk, or simply Dnepr.

The end of April in Ukraine epitomizes the start of the summer. Yes, it is still far from the muggy, sweaty, overbearingly hot July and August, and even from June, full of sunny hot days and yet refreshing cool nights. It is the time when nature fires up all jets of budding, blooming, blossoming... Boris and Georgiy are avid fishermen, but I am not, thus I joined this trip more as an observer; besides fishing, we were testing our newly synthesized mosquito repellent on ourselves.

Friday evening in April at Farther Lakes gets colder by the minute after the sun goes down. A wet twig in the bonfire shoots a bunch of sparks in the darkness above, and its loud crack breaks the idyllic silence.

I do not remember who went down to check on the fishing gear (we left it set for the night, a few yards down to the lake shore from our camp); I think it was Boris.

He returned quickly.

"Guys, the water level is down ... Way down!" He said with quite a concerned voice.

(Every resident of Dnepr and its vicinity knows exactly what this means: the dam shutters are put down, holding the river waters running south. This happens a few times every year during spring and summer, and when it happens, it indicates a major problem — contamination of the river with wastes that were "accidentally" dumped by one of the mega-plants or industrial giants, built on the banks of the Dnieper upstream. As a cheap source of water that is used for cooling, washing, soaking, and so on, the river was abused by these monsters, not rushing to admit the responsibility for contaminations or take steps to clean up their own mess, which usually manifested itself in the silver streaks of dead fish stretching for miles, the disturbingly bright-green algae *"soup"*, or on the contrary, the starkly brown murky waters.)

We looked at each other, concerned, and without saying anything rushed down to the shore. The water level was lowered by a yard, if not more — shiny, oil-black roots of the big birch tree that sat on the water edge were exposed to the air and resembled the tentacles of an underwater beast.

None of us had seen the level this low; it most definitely meant that the dam shutters were fully shut down for quite a while. Worried, we returned to the camp fire and for some time tried to filter out any comprehensible information from the old VEF radio that we had with us, endlessly tuning it in.

The Soviet radio stations were broadcasting typical bravura songs related to the upcoming May 1st Labor Day, and the only jot of anxiety that had seeded in our minds that breezy April night was a disturbed voice of some Swedish or Danish station anchor, repeating the word "Chernobyl" almost in every sentence.

You don't have to be a detective to figure out that something bad had happened at the Chernobyl nuclear plant, which sits right on the Pripyat river, feeding Dnieper above Kiev and above the Dneprodzerzhinsk dam, the one that smothered the river stream at Farther Lakes.

This was the first time when I had heard the name *Chernobyl* in association with the nuclear plant accident.

A sleepless night ended early. With the first ray of sun we packed, walked a few miles to the nearest village and took the first bus to the city. There, in the bus, we learned that there indeed was a fire at one of the Chernobyl reactor buildings; the fire was put down, but at a cost: two firemen lost their lives.

The situation has stabilized.

In recent years, I frequently contemplate a thought: how would the Chernobyl accident and its aftermath go, if the features common to our daily life today — internet, cell phones, blogging, social networks, and many, many others — had been available in 1986? That would include the speed and the volume of information sharing, news coverage, openness of the Soviet Union authorities (both internal and external), response of the world community and its actions (help and/or sanctions) and much, much more.

The answer is equally frustrating and gratifying.

Clearly, the magnitude of the catastrophe would have been much smaller, starting from the night of the accident on April 26th, 1986. Remember that the only available means of communication between the decision makers (plant management, nuclear experts, State and local authorities) and the plant staff (engineers, technicians, operators, etc.) was the phone, the good old land line. No mobile technology or cell phones available meant the loss of the precious minutes when the accident was in its initial phase. Dithering, procrastination, dragging in the decision-making process and the actions resulting from it were definitely affected by the information shortage. Internet, Skype, Facebook, Twitter would have ultimately made the information flow, exchange, and distribution much more abundant and rapid both in the first days and weeks of the aftermath and during the following months of the clean-up efforts. From my point of view, it would have had a history-changing impact not only on my personal involvement in the Chernobyl accident, but more so on the following key events related to it:

- Initial assessment of the situation after the Unit 4 failure and measures to prevent further deterioration of the reactor by both plant management and the personnel of the Unit 4 night shift;
- Full understanding by the State authorities and Communist Party leaders of the accident severity, related contamination levels around the plant and especially expanded contamination of the remote areas by radioactive plumes (including not only USSR but other European countries);
- Steps taken by the plant management, related industry officials, State and local governments to prevent further spread of radioactivity, clean-up and decontamination measures following the accident;
- Agonizingly painful story of Pripyat evacuation (about fifty thousand people were moved the next day after the accident);

– Political, social, moral impact of the accident to millions of people in the USSR and around the world.

One can only guess how the Chernobyl tale would have paved its way in the world's history, particularly, how this would have affected the fate of the existing and newly built nuclear power plants, for example, the Fukushima Daichi plant.

Ironically, about a year prior to the catastrophe, Anatoliy Majorets, then the Minister of Energy and Electrification of the USSR, decreed that information on any adverse effects caused by the functioning of the energy industry on employees, inhabitants and environment, were not suitable for publication by newspapers, radio or television. This decree had perhaps the most overwhelming effect on the available information related to the Chernobyl accident, even months after it happened, because it literally gagged everyone who worked in his Ministry's network. Imagine how this would have been possible in the era of internet.

But I digress.

How did we live during that incredibly hot and sunny summer of 1986?

No one can say that people were oblivious and unconcerned. In the very beginning, the absence of the official acknowledgement of the accident's real scale and grave danger it put the world in, and fully manicured *TASS*[1] daily news, made rich soil for a variety of rumors and gossip that were spreading like weeds.

In early May, when the Ukrainian capital was really close to a total evacuation, the most outrageous rumors scattered mightily in the country, one grimmer than the other. Strangely, it seemed that nothing was out of the ordinary in Kiev: the Labor Day parade was held on radiation-steeped Khreshchatyk, and multicolored wave of the Peace cycling peloton waved through the city roads ... Meanwhile, in Pripyat the families of plant workers were already evacuated in a hurry (including my cousin Peter's family, who at the time worked as an electrician at Chernobyl NPP — his younger son later had a lot of health issues related to the radiation over-exposure). Because our institute was

---

[1] (Russian: *Telegrafnoe Agentstvo Sovetskogo Soyuza*) Telegraph Agency of the Soviet Union.

closely related to the headquarters of the regional Administration of Civil Defence, so the officers from the *Military Department*[2] knew of course much more than the filtered information from the government sources to the media, and knew it first-hand; even a tiny bit of information that had seeped through that source gave all teachers and scientists of the institute a nightmarish feeling of helplessness and doom. The overall lack of reliable and trustworthy data about what happened and what was going on there, multiplied by the knowledge of the danger of radioactive contamination that my colleagues at the institute carried due to their professional involvements, was literally placing all of us in a state of shock. Here are some of the most memorable rumors and realities of those days.

- Kiev is drowning in the chaos and the full-scale abandonment, as in times of war.
- For a one-way train ticket — anywhere, just to get out of the doomed city — people pay up to a thousand rubles, and yet all ticket kiosks are closed.
- The main railway station and all airports are cordoned by the armed soldiers with vicious dogs.
- Mothers literally throw their children into the windows and doors of departing trains, just to get them out of the radioactivity-laden capital.
- Thousands of troops are entering Kiev to confront the riots and looting.
- A threat of a radionuclide seepage into the Dnieper is very real (in fact, the Dneprodzerzhinsk dam was fully shut down when we were fishing). Therefore, it is necessary to stock up drinking water, in the maximum possible amounts, since all three water-uptake stations of the city could be shut down any minute and will stay closed indefinitely.

---

[2] A department in charge of military (secondary) education in Soviet and Russian Institutions of Higher Learning, responsible for training the students to become the Army reserve officers (lower ranks).

– It is necessary to take whopping doses of potassium iodide, in case the radioactive cloud passes over the city and iodine radioactive isotope will accumulate in the thyroid gland.

On May 8, the Ukrainian Minister of Health Romanenko spoke to the public. He stated the necessary preventive measures: frequent hand washing, daily wet wiping of the apartment interior, careful removal of dirt from shoes, keeping the windows sealed. Nothing was said about the true state of the radiological situation in Kiev and in the republic — his report was obviously carefully doctored by the *Politburo*.[3] Meanwhile, consumers in all Dnepr city wet markets spontaneously boycotted the vegetables and fruits (particularly, potatoes and apples) sold from trucks with Kiev regional license plates.

Gorbachev appeared on television for the first time on May 15 — more than two weeks after the accident. Many westerners do not pay attention to this fact, but to me it was just as bad as an overwhelming majority of Gorby's internal political decisions and choices, or rather indecisiveness and spineless ways to handle the country that for centuries relied on a strong leader, be it a czar or a Party boss.

In his appeal to the Soviets, *Glasnost*[4] receives a huge kick in the keister: profoundly talking about complete control of the situation at Chernobyl NPP, about the role of the Politburo, and about the heroism of Soviet people, he did not drop a word about the true scale of the catastrophe, about the potential consequences, about the containment actions related to the Chernobyl accident. The Secretary General concluded with a standard psalm about the grave danger of the nuclear arsenals in foreign countries, which could cause the nuclear catastrophes thousands and thousands of times worse than the Chernobyl disaster.

---

[3] (Russian: *Politickeskoe Buro*) Short for Political Bureau, the executive committee of the Communist Party of the Soviet Union.
[4] The policy or practice of more open consultative government and wider dissemination of information in the former Soviet Union, initiated by leader Mikhail Gorbachev from 1985.

Mid-May onwards, radiation fear as a result of the information blockade had reached its climax. Rumors stated that the damaged reactor was ablaze, that hundreds of people were irradiated to death. The first eyewitness reports and stories emerged from liquidators: they came through rare phone calls, through letters, through relatives who had the guts to come and meet with loved ones, face to face in the vicinity of Chernobyl. News was not as scary as rumors, but was still in frightening dissonance with TV and newspapers.

Dnepr streets were abundantly and regularly sprinkled in the mornings, especially in the central areas, where the bigwigs lived. The early summer of Dnepr was irresistibly beautiful, and its flowery, refreshed boulevards and avenues drived up the anguish and suspense. Oddly, such splendor revived the visions of June 1941, right before the Nazi troops began the invasion of the USSR. I saw that similar beautiful summer morning in the movies; the serenity of Soviet border and nearby towns shattered in a matter of hours, and the brutal reality that swiftly replaced the peace and happiness was in stark contrast to those beautiful morning streets.

Two colonels from the Military Department were the first at the institute, who were called "for special military training". My colleagues and I had a rough idea of what sort of "training" they had to expect, particularly because such call of higher ranking officers, although not impossible, never happened before.

However, the reality, as it turned out, was far ahead of the assumptions.

After Gorbachev's speech, the information started slowly trickling through official channels. I saw the first TV footage from the NPP some time in late May or early June. Nothing special; a weird-looking heavy military bulldozer, soldiers in unusual outfits swarming around it; all this against the backdrop of a large industrial building. Polished phrases of the anchorman about the high morale of our troops. This short report (less than a minute) was followed by footage of the art festival "Kiev in the Spring". *"After a successful performance at the festival, The Ural folk choir is touring to various locations of the Army units involved in the clean-up efforts"* (later on in the aftermath, the choir members, who received substantial doses, tried for many years to obtain recognition of their diseases as related to the Chernobyl accident, but to no avail).

By the end of May, a massive mobilization of Ukrainian male population to the "special training camps" had begun, enlisting former soldiers and officers to the Army reserve — initially by hundreds, but soon by thousands. Since our institute was the only specialized institution of higher learning in Ukraine with a secondary (mandatory, mind you) military *chemical* training, everyone was anxious to expect that our personnel, teaching and research staff, would be drawn promptly. For some unknown reason (I actually learned this later, when I was already in Chernobyl), the first several waves of summoning swept through Kharkov and Donbass area. Dnepr and its vicinities were not affected massively until the fall of 1986. Nevertheless, the first of the civilians in our institute who went to the ChNPP was Pyotr Kondratiev, a technician, a colleague of mine from the Department of Corrosion.

The second one was me.

Actually, I volunteered.

To understand how this action was simultaneously bold, stupid, and impossible, the Western reader needs to keep in mind that I had a Doctorate degree and therefore was "whitelisted" (protected) from going to the special Army reserve camps by joint decree of the Ministry of Defense and Ministry of Education of the USSR. But I did not care. I wanted to go.

Until now, I cannot fully explain to myself what was the last push that sent me to meet *my Chernobyl*.

It was one of the sweat-drenching June evenings; I visited my mom and dad in their apartment, and was just about to leave, when the all-state TV news program "Time" aired a short report about the work of the heavy combat engineering vehicle *IMR*[5] at ChNPP. My late father, who at the time was still a very strong, healthy man, was watching that narrative with a luminous face. That did not draw my attention as it should've, and I left soon. Next day my dad, remembering his brave military past (he fought Nazis as a tank platoon commander in World

---

[5] (Russian: *Ingenernaya Mashina Razgrazhdeniya*) Combat Engineering Vehicle, a military engineering vehicle based on battle tank chassis; other accessories such as crane or back hoe can be attached.

War II from 1942 all the way to the victory in 1945), and being caught up in the alluring waves of patriotism, went to register as a reservist "*for Chernobyl*" at the District Army Enlistment Office. Naturally, he was denied: his age was way past all reasonable margins. He returned home very upset (per my mom), muttering something like "They don't know diddly-squat about that." Eventually, he gave up on this incredibly daring idea, but his gallant thrust somehow trapped me, too. I recalled a chat with a taxi driver that happened a couple of days earlier. One driver from his garage had returned from Chernobyl, where he was sent on a "*business trip*". (I learned later, that the drivers already were in an acute shortage at that time because of the frequent and excessive radiation exposure. Thus, everyone who had a professional driving license and above thirty years old, was hastily enrolled for such "business trips" on mandate of the local authorities and sent over to ChNPP.) This fella, said the cabbie, "came back all in blisters and with leukemia in acute stage". He blasphemed the radiation and claimed that the only viable option to fight it is drinking like there is no tomorrow, otherwise you are doomed. So I recalled that conversation and then thought that no matter how poorly we were taught the military chemistry, focused predominantly on the protection against weapons of mass destruction, no matter how laughable was the teaching competence of colonels in the Military Department, the essentials related to the effects of radiation on humans, the basics of the infamous brochure "Methods of protecting military personnel against the devastating action of nuclear weaponry" were welded deep in my manly brain. These tidbits had prevented me from drowning in a dark abyss of the unknown, the fear, which that driver was clearly experiencing over there in Chernobyl. I remembered the absorption coefficients of penetrating radiation for various materials — armor, cement, brick, leaded glass. I remembered the good old *depeshka,*[6] with its gravely subrange up to 200 roentgen[7]

---

[6] DP-5A, standard issue-type dosimeter of the Soviet Army in the 20th century
[7] Roentgen (R) is an old unit of measuring the radiation intensity. In the USSR industrial and military dosimetry, multiple devices were using this unit, either in roentgen per hour (R/h) or milliroentgen per hour (mR/h), applicable to so-called "penetrating radiation". It defines X-ray and gamma-ray intensity, but cannot precisely predict the body's response to the irradiation. Modern SI unit

per hour (if only I knew then, how soon and how often I would have to rely on this dreadful subrange!).

I remembered many things related to radiation and protection against it. I believed that I had the knowledge, or at least the basics of the knowledge, which gave me some chivalrous sense of potential utility. Literally from my first day at ChNPP I realized quickly and abruptly that such knowledge of mine turned out to be of little use, but at that time I was full of pride, awareness of my own importance and a noble desire to help the country and to protect the military and civilian personnel — I am mocking myself a bit, but thoughts like these were hiving in my head when I was marching to the Enlistment Office. These thoughts were also resting on a flimsy foundation of the traditional Russian belief in your own good luck and falsely adored credence of our natural ability to cheat the fate.

Anyway, in early June I voluntarily went to the Enlistment Office.

The young captain at the office entrance booth initially did not get the reason why I was there. He said that I had nothing to worry about, they did not call me for this duty and will not call, since I was *whitelisted*. But when I repeated that I wanted to go of my own free will, he gave me a sardonic look and pointed to the right room in the hallway, where a washed-out bald major retrieved my personal folder from the endless line of cabinets.

"Why the hell do you want to go there, are you tired of life?" He said inexpressibly.

I sort of froze for a moment, but then erupted in a long and pointless tirade, explaining my reasons. I figured that he couldn't care less after a couple of minutes in my spiel.

With the same stone face, while filling in the appropriate record card, he said that when the summoning arrives to me by mail, I would

---

of radiation intensity is Sievert (Si); the ratio between R and Si is 1 Si = 107.2 R. There are multiple ways of measuring and converting the units that are related to both radiation intensity and absorbed dose (or biologically effective dose); the author will use henceforth a R/h (mR/h) unit, which was applied to all radiation measurements in the Chernobyl accident and its aftermath in 1986.

have to return here again, pass the medical examination, get the money, travel documents, and *go there*.

My newly created card was placed in the folder, which again disappeared in the dreary line of file cabinets. The deal was done. Literally.

The following days of waiting were filled with a painful search of at least something, anything that would give me an idea of what to expect at the ChNPP, a description, even a hint of what the liquidators were doing over there, about their daily life — where do they sleep, eat, what do they do when not involved at the station work... My mom was mainly interested in the latter. I, however, had my share of sipping from the military cup in the past, so there was not a bit of illusion or hope in my mind.

However, in the following few weeks nothing really new about the work and the life of liquidators was revealed in the newspapers or on TV, to the extent that would help me or my relatives to get a better idea of what to expect.

In the meantime, in July 1986, my last pre-Chernobyl days were slowly coming to an end. The institute had already passed through the graduation mishmash. Life in the departments fell into that sleepy-relaxed summer schedule, when the teachers disappeared on leave until September, the graduate students moved to a *free work* schedule, and researchers, especially the females, spent more working time in the Nagornyi Camp wet market (conveniently, it was just outside the institute gates) than in the labs, snapping up cheap vegetables and fruits and simultaneously trying to identify with suspicion the trucks that had brought the produce from the near-Chernobyl areas. My personal life, which was tangled up to the limit that summer, was somewhat overshadowed by the upcoming leave; Laura, my future wife, was certainly worried about my nearing departure. However, being a stoic woman, she did not blink once and supported me as best as she could.

The summons came, as always, unexpectedly, when I began to wonder if I would be called at all. I was anxious for so many days and nights, but surprisingly, this simple message that was resting on the bottom of the newspaper pile in the mailbox somehow eased my worries.

A standard text, printed on a small piece of paper with a pale uneven font of the Enlistment Office typewriter said: "... is called upon to participate in the *special training camp exercises* for a period of up to six months ... to report to the District Army Enlistment Office for document procurement and medical ... within the next 48 hours".

Feeling somewhat lightheaded, I went to the institute *"to say the last goodbye"*. Department of Human Resources, represented by its Head, Comrade Naumov, instructed me with a traditionally sincere wish to *"keep up the good work."* The Accounting Department, where I had to set up a temporary leave, guided me with compassionate glances. If the cute accountants knew then, just how much troubles all liquidators from our institute would cause them, I believe they would have stopped us by any means.

In the long hallways of the Research Center everything was immobilized in the amber and dust of summer heat. The Polymer Laboratory, separated from the outside world by heavy curtains — protecting the inhabitants both from scorching sun and from potentially radioactive air — was filled with cigarette smoke that hung in layers up to the ceiling. Boris and Georgiy, who were resting at the corner sofa, drinking tea and smoking, met me with a bit of both curiosity and confusion. A few weeks ago, when I told them that I had volunteered for Chernobyl assignment at the Enlistment Office, they seemed to think that I was joking. But now the paper with pale carbon-copied letters has generated a proper respect.

An old record player, covered with solvent streaks, swayed the plumes of smoke with mellow tunes of Stevie Wonder. My fellows were relaxed and quiet. It was late evening, when the institute was almost empty. We drank the send-off shots one by one, snacking on seasonal flavors of the neighboring wet market. I, too, was inexplicably calm. Probably, I knew that my Mother Army had already spread its calloused but caring hand over my head, and whatever it was, for the next few weeks or months my fate was *"signed, sealed and delivered"*. Apparently, Boris and Georgiy did not think so, because they were exaggeratedly focusing on neutral topics: the fall trip of students to the collective

farms to help with vegetable harvesting, the latest gossip from the institute administration — funding, budgetary constraints, the football season beginning, and other small talk.

I realized the magnitude of my situation on the next day, when, after
a *complete medical examination* by the whole horde of *four* doctors that
lasted less than half an hour,[8] I left the walls of the Enlistment Office
and read the call of duty order.

"... leave for the location of the mil./est. No. XX979 ... to be assigned
a position of Deputy Commander of the Assessing-Analytical Station
... to replace Sen. Lieutenant Al. Holodov ..."

The paper was printed, not typewritten, with my particulars inserted
by hand. This meant that the circulation of it was supposed to be no
less than several thousand copies. *Maybe tens of thousands.*

This was perhaps the first time when I felt the magnitude of the liquidator engagement. A cold chain of goosebumps trickled down my
spine, despite the hot day.

Deputy Commander of the Assessing-Analytical Station ... I recall
how the station looked: a "Ural" truck with an attached trailer, fully
insulated with individual systems of air filtration (certainly, including
filters for radioactive particles), instrumentation, calculators, reference
books with an enormous number of tables and other data (computers
were not yet that common), walkie-talkies, field reports, calculations,
monitoring, reconnaissance. While studying at the Military Department (which was one full day every week of each semester), we were
trained for this duty, so I had a general idea of how to operate the station and how to lead the squad, which included up to twelve reports
in both the main truck and the trailer. What kind of work the station
was doing at the ChNPP, I had not the slightest idea of course, and
I was really afraid to mess things up at the beginning. However, as

---

[8] By the way, the wretched, deceitful role of multiple medical personnel along
the struggles of the liquidators before, during, and after their work at ChNPP
earns a separate book. Many of them were clearly instructed to conceal the
results of our medical exams, especially in the earlier months of the clean-
up work, the actual state of our health, the level of damage caused by the
radiation to our well-being, so that the real impact of the accident on the health
of liquidators will be hidden from the public.

always, my confidence was banked on the rigid pillars of the Army's infamous all-around simplicity and its ability to heave any wisdom in anyone's head, using a simple Soviet military concept: "Don't know how? — We'll teach and show; Don't want to learn? — We will force you regardless!" As much as I was mentally prepared for such emergency training, I was somewhat worried: as far as I remembered, the position of Dep. Commander was typically filled in by Captain-ranked officers, but I was a Senior Lieutenant, *starlei* in Russian slang. However, I was replacing a *starlei* as well, so I eventually laid this concern to rest.

Packing was done quickly. In the evening of the same day, at last, I was already standing on the platform of the railway station, breathing the last luxurious air of civilian life, while waiting for the departure of the train to Kiev. Laura walked with me only to the railway station doors; earlier, we had decided that theatrical scenes alike "*My dear one is going to the war*" would be too much.

In the four-bed train compartment there were two other guys who, from their angle, were unfortunate to get called to Chernobyl as the reservists. The fourth passenger, their colleague, defected and did not show up at the station; this was not a rare instance, as many men exploited any possible excuse to desert, literally going a long mile for this. A couple of years later, when the number of available reservists substantially shrank, the Army used all available measures to fill in the quota, including raids to the houses and places of business, but still, there were incredibly clever individuals who escaped this call of duty. The reason why the actual reserve availability was not as high as the theoretically available was first of all the age limit: only those who were 30 years and older were eligible to be the reservist liquidators. Of course, as with anything related to the Army, there were *whitelisted* guys of all sorts including doctorate degree holders, like me; there were health-ineligible men (in my beloved former country, a hefty bribe to the risk-taking psychiatrist would have conveniently produced such a clearance), there were sons, cousins, nephews and other relatives of the local Party and establishment bosses... All in all, there were tons of guys who escaped the duty call without any punishment.

The two fellow passengers turned out to be the drivers. They methodically emptied two bottles of vodka over a span of the evening, drowning their stress and fear of the unknown in alcohol. I smoked several cigarettes, sitting at the open window in the hallway, and tried to shake off the gloomy thoughts about what was approaching me with the speed of this express train and trying to imagine what Holodov looked like, how skillfully he managed the data processing while commanding the work of Assessing-Analytical Station, the AAS. When the plexiglass map of the area, vertically bolted on the table of the AAS trailer, suddenly blossomed with tiny cartoonish nuclear explosions, I realized that I was falling asleep. The rest of the night starlei Holodov monotonously explained to me *the principles of the method for calculating the individual irradiation doses during the passage of a motorized brigade through the epicenter of a nuclear explosion*, occasionally interrupting his explanations by shrieking: "Why the hell do you want to go there, are you tired of life?"

... I woke up because of a shove to the back.

"Belaya Tserkov[9] ... The stop is ten minutes, hurry!" One of my travel companions shouted.

The train arrived at the first rays of the sun. The three of us stepped on the platform, squinting after poor sleep. We talked to the station supervisor, asking how to get to *perevalka*.[10] As it turned out, it was about an hour-long bus ride away from the station, somewhere on the outskirts of Belaya Tserkov. The bus goes there once every three hours, starting at 11 am.

Learning that we were heading to *"the war"* — a lot of locals in the areas adjacent to Chernobyl called the accident and the aftermath this way — the super instantly mellowed and took us to a miniature station restaurant, where we were offered hot scrambled eggs with sausage. My companions clanked the glass and the vodka bottle, tempting me, but I did not feel like drinking this early, and limited myself to a cold black tea.

---

[9] A small town in Kiev suburbs.
[10] The Army base where reservists have to exchange their civilian clothing for the military one and then get a transport to their destinations.

The morning dragged on and on like a limping turtle. I was tired of waiting, tired of smoking, tired of pacing up and down the platform, but most of all tired of uncertainty. The first bus came by with a mere half an hour delay. When we finally jumped off the bus at the checkpoint of the Army base, the morning's scrambled eggs were persistently looking for the company of something more significant — *borsch*,[11] meat, potatoes...

The checkpoint guards barely looked at our documents and sent us to *perevalka*, a dull three-story building, tucked in the corner of the base's monumental brick fence — I imagined that on top of it there should be a walkway, like on the Great Wall. Barred windows of *perevalka* immediately pulled up an analogy with the prison. As always, the Army was all about the caste: I was sent to the first floor, and my companions, since they were the *mere sergeants*, to the third.

I received a field uniform (a collared jacket and for some unknown reason the breeches as a pairing piece; breeches were not a regulation uniform for ages, and I was stunned that these were actually brand new!), a camouflage-green flat top officer's hat with cockade, a reefing jacket, a military backpack, a pair of boots (not chrome, but a soldier's *kirza*[12] — what a drag for an officer!), a belt, a pair of underwear, and some other unimpressive Army attributes. I dropped in the backpack a few things that I took with me from *pre-Chernobyl life* (very soon this will become one of the significant, if not the most important, markers of my lifetime), and shoved the civilian clothes in the sports bag that I arrived with. A heavily breathing fat *prapor*,[13] using perfect penmanship, entered my name in the register and took my bag for safekeeping in the impressively large storage facility containing the civilian things of countless others. I was certain that the *perevalka* inside is much bigger than outside, due to some secret Army trick with time-space continuum.

---

[11] A famous Ukrainian soup.
[12] The type of artificial leather that imitates the pig skin; the material is mainly used in production of soldier's military boots.
[13] Chief Warrant Officer.

In a poorly lit, stuffy storage facility with the single barred small window above the *prapor*'s desk, among hundreds and hundreds of suitcases, bags, backpacks, coffers and other things on the shelves (some were really strange, even bizarre — I vividly remember a bicycle in the corner and three folded Uzbek *chalats*,[14] tucked in between the bags), dressed in the poorly fitting new uniform, I felt readied to get convoyed to Siberia.

It was much nicer outside, for sure. I returned to the checkpoint, received an order to arrive to the barracks of the 8th Company and to be on standby *"until requested"*.

My brand-new boots were not broken in yet, and therefore walking across a huge camp, almost to the end of it, to reach the barracks took me nearly forty minutes and gave me a pair of quick blisters.

The barracks were empty, except for the sergeant-on-duty, sleepy and bored. I went inside and sat down on a stool in the reading room, not risking to disrupt the sacred silence of the large barracks with rows of two-tiered bunk-beds. The deceitfully dry heat of late July was funneling in the room through the half-open window, mixed with the chirping of a sparrow flock that occupied a huge poplar next to the entrance. A loud commander's voice scolded the subordinate somewhere in the distance, not withholding the frequent mentioning of the subordinate's mama. The air in the room was heavily saturated with the smell of the soldier's cheap shoe shine cream.

The sense of belonging to the apotheotic Army machine became so real that for a moment I truly panicked.

I was despondent. The ChNPP, the AAS, and Sen. Lt. Al. Holodov were so distanced from this, that I almost disbelieved in their existence, and instead had strongly accepted the heavy odors of the barracks, push-ups in front of the barking drill sergeant, savagery of the platoon bullies, *political information seminar* for the whole platoon twice a week before breakfast.

The sheer Army essence had started overtaking me. The boots suddenly became huge and exuded a hefty smell of tar. The belt was strangling the waist. The jacket was loose in the collar — my neck seemed

---

[14] The color-striped cotton wool coats, typical for rural areas of Central Asia.

to float hopelessly between its banks. I was starving, and the sense of hunger became even greater when I heard through the window the roar of soldiers' vocal cords, happily yelling the marching song *"Don't weep, my girl, the rains will pass"*. I looked through the glass; a platoon was cheerfully stomping to the building next door, most likely the canteen. Only two places at the Army conjure such joy in soldiers' hearts: baths and canteen, and since I didn't see the towels in their hands, I rightfully concluded that this building must be a canteen.

Saliva filled my mouth, but I suppressed the hunger and courageously decided to wait for that very *"until requested"*.

The whole eternity seemed to vanish, and I was desperate. The sergeant-on-duty, who passed by the room, looked at me pitifully, grunted and said that I could go and get something to eat, since *the dispatch*[15] will be later, maybe even in the wee hours: in addition to several small teams coming in tonight, there will be a full charter train of reservists arriving from afar; only then everybody will be wheeled to the brigade on *borts*.[16]

This was the first time that I learned that I would join the forces of the infamous 25th Brigade of Chemical Defense, which was involved in the dosimetry reconnaissance, decontamination of ChNPP equipment and territory, and collection and disposal of radioactive waste almost from the first day of the accident aftermath.

The Sergeant also said that the Brigade is about three hours' drive away, and since we will arrive there in the early morning, it's best for me to get some sleep in the afternoon — I can pick any bunk in the barracks that I want, since *the regulars* are out in the field for another week.

Lunch in the canteen was rejoiced by an incredible number of flies and a split pea puree, a menu staple of our beloved Army. Since the prospect of dinner was not on the horizon, I stuffed myself up to the brim. Flies and the Army are touchingly inseparable. I dwelt on this concept for a while, flushing down the puree with a suspiciously cloudy fruit compote.

---

[15] The release of reservists to their military units.
[16] Tented trucks with side-to-side wooden benches inside; widely used to carry the troop companies up to 30 soldiers.

My mood changed for the better, and the empty barracks now looked inviting, considering the limited amount of sleep that I had last night. I gladly took off the belt, boots and hat and stretched on the lower bunk in the far corner of the barracks. The remaining time was filled with the influx of more and more reservists. Soon the barracks were packed to capacity, but the charter train had not yet arrived. The air was gradually filled with the traditional odors of food, sweat, shoe polish, new leather... I thought that the old barracks had finally relaxed, receiving a full portion of human stuffing inside, without which it felt lonely and grumpy.

Laughter, chats, anecdotes, cursing merged into a monotonous noise, hypnotizing my tired brain. I dozed off. The sound of multiple truck engines outside the windows woke me up after some time. Finally, the reservists from the charter train had arrived. The whole train was from Russia, as we all soon found out, drawn up outside the barracks in the uneven and long lines. It was about ten in the evening. Some brass-voiced commander tried to brainwash us in a hurry for the last time (typical Party-approved lines: " ... your duty is to stop the spreading of the nuclear threat ... the Motherland counts on your heroic efforts ... " " ... you all are the best that this country has in order to do this job ... ").

A minute later, we were rolling toward the *"atomic plague"*, as someone in my *bort* joked.

We sat on benches, six on each; they had no backrests, and the leg space was definitely not the primary concern for those who assembled the benches. The tent had no windows, only a wide opening along the back, and the air inside immediately got stuffy. The neighbor smiled at me, offering to have a drink from his bottle as a welcome gesture. Suspecting that the content in it was not lemonade, I carefully took a small sip, and the fumes of poorly distilled hooch hit my nostrils, throat and very soon the head. A simple snack was a piece of bread and a garlic clove (the latter would for a long time become the "best friend of the liquidator", from the home-sent parcel containing several cloves, and even a sheaf; garlic was believed to be a sure help in "decontaminating" the body). Hooch eagerly filled the brain. Somebody smoked, despite the strict ban given before leaving.

Belligerence of the brew, combined with cigarette plumes and a long ride in the dark sitting backwards, did not agree with my stomach. The stress of the past day rolled there in nauseating waves, but I drove it away, thinking of all that I left behind — my family, the institute, the work, the stuff that I *"cooked"* (synthesized, that is) in the lab just before my departure.

The *bort* suddenly slowed, veered left, then rolled over a small hump down from the highway, and a few moments later stopped.

"Everybody out, on the double!" Someone yelled outside the tent.

Following the others, I jumped into the pitch-black night that was sparingly sliced by the beams of passing cars. A vast flat area, covered with gravel, was surrounded by tall forest, barely recognizable from the distance. No one hurried to draw up the line; confused, we clumped together, not knowing what to expect but still trying to figure out something in the dense darkness that had swallowed the long line of *borts* and all of us. We instinctively clung to the trucks as if they could protect us or at least return a sense of reality.

Out of nowhere appeared a ghost-like figure, dressed in white clothes, with white shoes. It ran toward us, waving its hands in the air; the voice of the phantom barely broke through a white towel, tightly wrapped around the lower part of its face.

"Guys, cover your mouth and nose any way you can, do not inhale the radiation!" The ghost yelled in panic.

Many hastily took out handkerchiefs, spare foot-cloths, underwear, covering their faces. I look with suspicion at the phantom.

"Suckers, welcome to the Brigade!" The towel on the ghost's face muffled his laugh.

"Savich, drop it this instant! If you continue scaring *zamena*,[17] I'll chain you down in Brigade, and you'll sit tight until the snowfall starts!" Someone right next to our *bort* barked distinctly and sternly.

The white figure disappeared in the night.

I checked my watch. 1:20 am, July 31, 1986.

My Chernobyl has begun.

---

[17] The replacement personnel.

PART 2

Forewarned…

July 31, 1986
Chernobyl NPP, first hodka[1]

> *"The primary and the most important task of dealing with the ChNPP accident aftermath was to clean up the areas near the destroyed unit No. 4 from highly radioactive fragments of nuclear core ejected from the reactor, which involved collecting, transporting, and burial of such highly emitting fragments. Early on, at the beginning of May 1986, on the territory of the Industrial Zone, due to the timely and efficient removal of the most active fragments, the levels of gamma radiation were reduced by 10–20 times, which allowed to prepare a wider front for subsequent decontamination."*
> (From the official Ukrainian government website dedicated to the ChNPP shutdown.)

The sound of trampling feet behind the thin tent wall woke me up. The sun drew a big dark cross, shadow from window framing, on my chest that was covered with a white sheet. Emerging vivid associations forced me to jump off the bunk bed, and I painfully rammed my head into the upper bunk.

The tent was empty.

My bed was next to the entrance, on the right, past a short two-door portal. "Doors" were just the roll-up double-sided curtains, and the tent had no floor. Alex Holodov recommended me this bed: nights here were very hot, and the first bunk was strategically placed near the window. Another privilege of the officer: the bed above me stayed vacant. Plus, I had a personal nightstand. And a personal stool. I felt like I was moved to the Ritz-Carlton. I might splurge with a hefty room service …

---

[1] The "jump", or working shift of a liquidator at ChNPP; varied from as short as under a minute to as long as 8–10 hours, depending on the radiation level and exposure at the working site.

The sandy floor cooled the feet. I looked at the three small blisters that I acquired the day before, then at the foot-cloths, wrapped around the boots for drying overnight. This was one of many enigmas of the Soviet Army: why almost at the end of the twentieth century did our soldiers have to be punished by enforcing the use of foot-cloths? If there was a shortage of socks in the country, I may get that, but no, generations after generations of military men had to learn, the hard way through blisters and blood, how to properly wrap them around the feet, before the pain from slid and bundled foot-cloths in the boots stopped torturing the defender of the sweet motherland.

The night before, Alex brought me here. He pointed with a flashlight to the bunk bed. "The sheets are not fresh, but I only slept on them a couple of times," he said laconically. I didn't care at all since I was half dead from fatigue and stress. He held the flashlight for me while I prepped the bed. Once I was done, he wished me a good night and left. The orchestra-like waves of powerful snoring, coming out of over two dozen youthful lungs, were rolling from end to end of the tent. Normally, I couldn't fall asleep while distracted by snoring, but after a spontaneously initiated briefing given to me by Alex and another officer, Igor (Chief Commander of the AAS and formally my immediate commander), in the half-lit battalion Command Tent, I couldn't care less about the snoring and dozed off the second my head touched the pillow.

Holodov was tall, thin, slouchy, with short-cut blond hair; he was wearing a well-beaten field uniform with camouflaged Senior Lieutenant's stars on the shoulder boards. Initially, I found him somewhat dry, a far cry from a great conversationalist. Igor, too, seemed to be not a totally joyous fellow. He was a direct opposite to Alex: short, dark-haired, with sparkly eyes that sat deep in the sockets, which made him look somewhat demonic in the semi-darkness of the Command Tent. For an Army officer, he was wearing a strange navy-blue uniform — trousers and jacket. The jacket had self-made boards with four Captain's stars on each shoulder; I learned later that such uniforms were worn by the ChNPP personnel. Throughout the whole night and morning, their faces carried an expression of extreme fatigue that borders on apathy. Later, I learned on my own that it was indeed a result of the regular continuous over-exposure to radiation *on lower fields*, areas with lower levels of radioactivity.

Igor and Alex took turns to talk to me, sharing a massive amount of specific information: while one was talking, the other was resting. This non-stop piling of data, chaotic and densely interlaced with specific slang, was discouraging. From the very first minute of this briefing I realized how little I knew about all those "what, how, when, where" related to their — and now my — job assignment. Their talk so far evoked only the dreary sensation of my own insularity. I shyly inquired as to when I would see the AAS. Igor looked at Holodov, shook his head, then silently got up and went out.

"Don't you worry about AAS, it won't go anywhere," Holodov replied with arcane notes in his voice; he waved his hand, pointing somewhere behind my back.

(We were sitting in the Command Tent, lavishly trimmed from the inside with a white cloth inlay. The Battalion of Chemical Reconnaissance of the 25th Brigade of Chemical Defense was located at the very edge of the camp. The Battalion Command Tent was located in the third, last row of tents, followed by the food block. When we finished the briefing, I went out and looked at the premises: a bit further, behind the barbed wire surrounding the car park, I recognized the dark shadows of *"soap dishes"*, *BRDM-2rh*,[2] and some other cars and trucks, among which, in theory, should have been parked the AAS.)

The long day, full of tiring events, took a hard toll on my brain, which resisted cooperating very soon into the briefing, but they continued to pour onto me the endless *"allowed daily dose"*, *"otsidka"*, *"square-cluster detection and recording"*. Then there were more colorful *"Moldovan checkpoint"*, *"Lelev PuSO"*, and a bunch of local geographical names — Dityatki, Kopachi, Chistogalovka, Oranoe... They both called ChNPP *"the Station"*, and since then I used that word as well: short and respectful, as it was meant to be. Closer to three in the morning, they finally got tired and stopped the torture. Learning that I had not eaten since lunchtime, Alex stepped out and returned, holding a can of spam, half a loaf of rye bread and a bottle of mineral water. Using

---

[2] Armored Vehicles of Chemical Reconnaissance.

Igor's jack knife as a fork, I swallowed all that in a minute, thinking how little a man needs to be happy.

...

"Breakfast time!" shouted Holodov, coming inside the tent.

With a gesture of a circus illusionist, he pulled from his pocket three pairs of cotton socks, still with factory tags.

"Forget about the foot-cloths. They are useless at the station, socks are more convenient, and distributed there by the thousands. Disposable, like surgical gloves," he said.

(I always recall this touching episode in relation to a scene from the movie "Forrest Gump", where Lieutenant Dan instructs Forrest and Bubba who had just arrived to Vietnam: "There are two things you need to remember while here: first, don't do anything stupid, like get yourself killed, and second, keep your feet dry!" Sometimes, the Army's wisdom is really, really inspiring ...)

After a quick face wash, I jogged to the officer's canteen of the Food Block for breakfast, which included oatmeal with canned fish, albeit good ("Saira!" — as I was told with pride by the dispensing staff), bread, butter and a pretty decent piece of processed cheese; the feast concluded with hot tea, which was recognizable as such only by the color.

After breakfast, Holodov and I rushed to the battalion operative meeting.

"For today's *hodka*, you will go as my *stager*,[3]" Alex talked to me on the go, breathing somewhat heavily. "This is my last jump; I will fulfil my quota today. Remember one thing: wherever we go in the Station, you follow me step by step, no deviations, no lagging. Otherwise ... " He pouted. "Once the operative meeting is over, we collect the team and roll to the Station."

In the Command Tent, there were several officers gathered around a simple wooden table on scaffolds. I noticed two senior officers — by age, not by rank —, the rest are reservists, mostly Lieutenants. I suddenly realized that I did not attach the rank stars to the shoulder boards. Darn.

---

[3] Understudy, intern.

A short, stocky, red-faced Captain in tight but well-ironed uniform read something from his notebook; since he mentioned a lot of geographical names followed by the personnel names, I figured that he wasdeclaring the dosimetry reconnaissance routes and the team commanders taking those routes.

"Captain Zelenov, battalion *COS*,[4]" Alex whispered.

"Holodov, you take a team of fifteen, today you go to HOYAT,[5]" Zelenov turned his head to us, as if he overheard Alex talking.

In Russian, this abbreviation sounded very close to a popular curse word, and since I was unfamiliar with the actual Station slang, I squeezed out a short giggle, but not aloud. Alex's gaze bolted me to the floor.

"Yes Sir, Captain!" Alex replied.

"A civilian engineer from *US-605*[6] should pick you all up at *razvod*,[7]" Zelenov continued. "The task will be given directly there, but most probably it's gonna be contaminated soil removal, where at — I don't know, and Brigade HQ doesn't know either. Whose dosimetrist is going today with Holodov? First Company? Second? Come on, sh*t heads!" He snarls. "If you won't figure it out on time, who would?! I have other cr*p to shovel today, and plenty!" He menacingly circled the group with a vicious gaze.

"Sir, today's shift is served by the First Company, dosimetrist Sergeant Gorin," a hand was raised to the left of Zelenov.

"That's better. Petrov, give Holodov his roster for today," the Captain's voice softened. "Alright, looks like that's it. Any hospitalizations? *Zamena*?" He turned to me for the first time.

"Newbie? *Zamena* for Holodov? What's your name? Belov, Be ... Belyakov, right?" He silently stared at me for a few moments, studying.

---

[4] Chief of Staff.
[5] (Russian: *Hranilische Othodov YAdernogo Topliva*) Spent fuel depository.
[6] (Russian: *Upravlenie Stroitel'stvom*) Construction Department, 605 Office, the head civilian organization for clean-up and containment work during the aftermath.
[7] Personnel line-up at the former Reactor 5 and 6 construction site at ChNPP; here, a daily distribution of liquidator work force had taken place in the summer of 1986.

I'm not a physiognomist, but I've always believed and continue to believe that your relationships with a person are largely determined by the first eye contact. Well, it appeared that I didn't make a good impression there.

"Look, this is your first *hodka*, just don't soil yourself out there," the Captain talked to me softly, but expressly. "Follow Holodov as if he is your mama duck, suck his tit if needed. He'll show you the ropes; learn and absorb everything really well, cos tomorrow, when he's gone, there will be no one to wipe the fish tails off your nose. Got it?! Dismissed!"

...

Our team consisted of older, aged reservists. Gorin, the dosimetrist, was anxiously flipping the switch on DP-5A, dosimeter, *depeshka*.

"Battery is messed up," he frowned. "There are no more spares anywhere in the whole battalion. Supplies Chief promises delivery every day, but... This one is almost gone! If *depeshka* stops working while we are at the Station, then what?" Gorin was clearly desperate.

I looked at the instrument. The dosimeter was worn out to the limit. The probe was attached to the rod with blue duct tape, the lock on the casing was ripped off. I had nothing to say, just shook my head sympathetically.

The Brigade dispatch area was located across the highway from the quarter's location. Building it on the other side of the road was not a clever idea by far, because one had to gamble, often quite precariously, while crossing the never-ending flow of vehicles that rushed to the Station; cement "mixers", military vehicles, cargo trucks, buses, *borts*, and all sorts of passenger cars were flying on the highway pretty much bumper to bumper. The highway surface was re-patched and covered over with asphalt multiple times, roadsides were frequently watered by *ARS*.[8] Both measures aimed to lower the spread of radioactive contamination by the dust from vehicle tires, which initially was an underestimated problem. Our team crossed the highway safely, in front of the

---

[8] (Russian: *Avtomaticheskaya Razlivochnaya Stantsiya*) Automated Distribution Vehicle, a multi-task car containing a cistern filled with either water or decontamination solution, and various spraying equipment, ranging from hand-held nozzles and sprinkling mechanical brushes to vehicle-mounted sprinklers.

BRDM convoy that moved to the main road from our battalion's car park — they were heading to the Exclusion Zone for radiation reconnaissance scheduled for every morning. Alex raised his hand, thanking the driver of the leading armored vehicle, which slowed down to let us cross the road safely.

The heat got worse by the minute. While the team was trying to find some shade close to the half-dozen cars that were melting under the sun, Alex was angrily arguing about something with a Captain, who apparently was leading the dispatch. I overheard fragments of their heated conversation: "... what the f*ck!", "... I was told it was its last time ...", "... this piece of sh*t already has to be buried ...". It seemed that something was not working out. Alex saluted the Captain with exaggerated precision and then walked toward an old *bort*; I followed my partner. His eyes squinted in frustration. Approaching, I heard he asked the driver waiting in the cabin: "On the way back, can you find a detour bypassing *Moldovans*?" The guy nodded affirmatively. This whole sequence looked way too strange to me; however, I kept quiet for the time being. Alex commanded us to load the *bort*, we climbed on without much enthusiasm. A few minutes later we were rolling to the Station. Vasily, the driver, softly whistled something unrecognizably melancholic. I was squeezed in the cabin between him and Holodov. Alex kept to himself, only at times cursing through his teeth. There clearly was a problem, but I did not want to pester him with questions — when it was necessary, he would explain.

Vasily changed the gear, painfully knocking my left knee with a shift stick. We were approaching the checkpoint. A simple slingshot bar, painted in red and white stripes, blocked the road coming from the Station, but the traffic heading there, on our side, barely slowed down. The actual checkpoint was just an old, sun-faded green tent. Two policemen in bulletproof vests, armed with AKM-47's, stood nearby, watching the traffic going to the Station. They had green *R-2*[9] hanging on the necks, but it seemed that they were having such "protection" just to keep up with the formalities. In front of the gate bar, coming from the Station,

---

[9] Polyurethane-foamed respirator, aimed to prevent dust particles penetration.

parked a truck; the dosimetrist in military uniform was checking the radiation levels on the car, sticking the probe at multiple places under the frame and wheel boots. Behind this truck I see a long line of "mixers", *borts*, buses, cars, *BRDM*, waiting to get checked. A rudimentary barbed wire fence, about six feet high, extended from the checkpoint, cutting through the dense forest on both sides of the road. The fence posts were already tilting, some really badly.

We were in the Zone.

I was a bit puzzled; I was expecting the fence and overall security of the zone to be more rigorous. But here... we have a primitive faded green tent, where, apparently, policemen and dosimetrists hid from bad weather, a flimsy gate, and a loose net of barbed wire as a fence?

We continued riding into the Zone. About fifteen minutes later, a wide asphalt-laid area emerged in a massive forestry on the highway side; the purpose of this area became instantly clear to me as soon as I saw the three decontamination chemists in *OZK*[10] gear and gas masks. They were washing the truck with brushes on extended rods. A pair of *ARS* nearby buzzed heavily, pumping decontamination solutions through a bunch of hoses, which were feeding the brushes.

Decontamination site.

"*PuSO-3*, Dityatki village. There will be two more *PuSO*,[11] at Zalesskoe and Lelev," Alex pointed out.

During my military education, I had the experience with "deploying the decontamination site", "deactivating the equipment and machinery", but that was purely a game, although with a bit of an Army-infused taste. Here, the gravity of these men's actions was far too obvious to me. This was not a game, I had not a shadow of a doubt. Feeling excited, I longed for action.

---

[10] (Russian: *Obschevoiskovoi Zaschitnyi Costyum*) All-Army Protective Costume, designed to guard the personnel from chemical and biological warfare; it covers the whole body and consists of a rubber-impregnated impenetrable coat that converts into overall, over-the-knee boots, heavy gloves, and a hood, paired with the gas mask.

[11] (Russian: *Punkt Spetsial'noi Obrabotki*) Area of Special Treatment, a decontamination and deactivation area that was treating contaminated vehicles inside and at the exit from the Exclusion Zone.

Vasily explained why *PuSO* is simultaneously a friend and a foe for us.

"The big deal with *PuSO-3* is that it is the last, therefore the most stringent contamination checkpoint at the exit from the Zone, and if your vehicle *"shines"*[12] on its tires or under the car body over 5 milliroentgens per hour, then you must go back to Dityatki *PuSO-3* for decontamination, where military chemists will wash the lower part of the vehicle with decontaminating solutions, and even — when needed — sand-blast the frame and body to remove the radioactive dirt and particles that were collected at the Station premises."

I was amazed that the *PuSO-3* was built so far back on the high-way from the checkpoint. But then I realized that with such high daily volume of traffic going out of the Zone, and passing through the check-point and *PuSO-3*, the amount of washed radionuclides must be really astronomical, and it would be dangerous to keep this clean-up facility close to the Zone border.

"And this piece of junk already sucked up so much that its rear axle emits over a hundred milliroentgens!" Vasily angrily punched the steer-ing wheel. "Yet the whole frame was sand-blasted so many times that I can see my reflection when I look at it; there is not a spot still painted there! It is all metal, it reeks neutrons!"

My jaw dropped.

"The deal is that if the car, having three consecutive decontamina-tion rounds at *PuSO* during return from the Zone — with sand-blasters, with decontamination solutions — still emits over five milliroentgens on tires and the frame at the exit checkpoint from the Zone, in theory it should be dumped in *mogilnik,*[13]" Alex said. "I saw a few fire trucks in one of them, maybe *the first ones*, you know, from the night of the accident in April..." He scowled, then continued.

"Well, get this: the guys on the checkpoint will send the *bort* to *mogilnik*, which seems to be proper. However, how is my team sup-posed to get back to the Brigade? There is no system designed for these

---

[12] Emits radiation.
[13] The burial ground, containment site, where the contaminated equipment, soil, and trash are compounded and kept indefinitely.

situations — you are on your own. You can try to catch a ride with any-one who goes by the Brigade, which may not be the case because there is a high volume of liquidators positioned before Oranoe village, where our Brigade is stationed. Besides, people are different — some take pity and get you on board, some are afraid that your clothing is still *shining*, so they just roll by. Two days back, after our *bort* was arrested due to the high emission levels, we spent about an hour and a half, standing in scorching heat trying to catch a ride — not a single bastard stopped, not even ours, the military, forget the civilians. Finally, the policemen shuttle bus stopped and picked us up; it took over three hours total for a trip back from the Station! And what did I get in return? The Supplies Chief squealed on me to the Battalion Commander that I gave up the counted *bort* for *mogilnik*. Think about it: *dozers*[14] have all the rights to confiscate and contain the contaminated vehicle, yet the Comman-der desperately needs a vehicle tomorrow for sending the teams to the Station, and there are no more clean ones! All others are contaminated up the wazoo," Alex grinned.

I couldn't make sense of this.

"Have you seen the license plate of this junk? MM 00-02, MM stands for *Mechta Mogilnika* ("*Mogilnik* Dream")," Vasily smiled sadly.

I looked out the window at the landscape, which for a long time would become so close and so familiar for me. Beautiful Polesye region villages were rolling one after another outside. The complete absence of people seems eerie, as in the post apocalypse movies. Doors and win-dows of many houses were sealed with planks and boards that were nailed criss-cross.

Fruit trees in the gardens along the road were densely ("*ryasno*", as they say in Ukraine) covered with monstrous size apples, pears, plums — the branches hanging way too low and literally cracking under the weight of the fruits. Small radiation fields strongly stimulated the cells' growth, including fruits and vegetables. Puddles from constant gener-ous watering of the roadside do not dry out even in the 30-degree heat.

---

[14] Dosimetrists.

Every five hundred meters along the highway I see posted signs: "DAN-GER! CONTAMINATED! —RADIATION HAZARD!"

Alex pulled a small paper package out of the gas mask bag on his side and gave it to me; it looked like a surgical patch or a bandage.

"Respirator, "*lepestok*",[15] civilian. Ever used it before?" He asked.

I was familiar with it, but had not worn it before. He showed me how to use it — nothing tricky. Clamp on the nose helped to fit the mask tighter; however, it was uncomfortable, as it put pressure on the nose and made it harder to breathe.

"Don't worry, you'll get used to it. The first commandment at the Station is to wear it at all times; removing it is discouraged, except in extreme cases," Holodov chuckled.

Vasily took an R-2 respirator out of his pocket. Alex gave him a dis-approving look, and I understood why. *Lepestok* is a disposable respi-rator; once you use it, you throw it away, no potential hazard. However, the polyurethane foam of R-2 is a liability, because it will filter the air from radioactive dust well, but the radionuclides will accumulate in the respirator outer layer. The pocket was not the best place to store such a thing, and Alex reminded the driver about it.

The highway splits; the right road goes to Yampol. The traffic to the left, towards the Station, was much busier. We were slowing down. Alex checked the time: *razvod* at the Station was carried out near the second construction site of the ChNPP, reactor units 5 and 6, which were close to completion when an accident occurred. It was imperative to be there before 8 am for the procedure of *razvod*. Besides getting directives from the "customers" (representatives of civilian organizations that used the liquidator work forces on the Station for various clean-up tasks), the teams received the mobile supporting equipment — cranes, scrapers, bulldozers and so on — most of which plainly couldn't leave the Sta-tion because of their heavy contamination. The traffic that morning was quite slow, and Alex pressured Vasily to go faster, although there was nothing much that he could do on a one-lane highway.

The conundrum of dual control was the plague of the system, explained Alex: on one hand, the team and its leader must obey the

---

[15] "Petal", simple disposable respirator, protects from inhaling dust and aerosols.

instructions of military commanders, but on the other hand, the civilian "*customers*" from multiple construction organizations and institutions often gave orders that were in complete contradiction to what the commanders said, and *vice versa*.

We finally arrived at the second construction site. I saw ahead the cyclopean concrete-laid cooling towers of blocks 5 and 6. We headed to our spot following the directions of a sweaty, stressed out Captain who appeared to be Commander-on-duty of the day: the teams were lined up around the *razvod* site perimeter, in front of the respective equipment. We had no extra gear for the day, and our troops were drawn up in a single file near MM-00-02. While Alex was reporting our arrival, we bravely sweated out in the sweltering sun. My face under *lepestok* quickly became wet — either from sweat, or from heavy breathing — its moisture saturated the textile, and it started emanating quite strong medicine-like odor. I regretted that I did not take off a long-sleeved shirt that had to be worn under the field uniform; in the heat, the jacket did not soak in the sweat and was totally non-permeable, producing generous tributaries of sweat that were running along the chest, sides and spine, spurred by a sense of imminent danger. I was not sure if this sense was caused by a real threat, or just by my imagination. The Station was barely visible from there, although the striped deaerator stack on the roof between the destroyed unit 4 and unit 3 was quite visible.

The *razvod* site was gradually emptied. Teams were sent to the Station, one by one. Alex was slowly returning to the team. It turned out that our customer had not shown up yet.

We were sweating seconds and minutes out. No one was allowed to leave or to hide in the shade. Ten minutes passed. Our troops were unhappy, but did not express it openly.

"Does this happen often?" I asked Alex.

"Sometimes," he answered, without looking at me.

A few minutes later, he cursed and ran to a group of superiors, who seemed to be about to leave the site. After another round of heated talk and yelling, we triumphantly rode on our *bort*, an infamous MM 00-02, to the Station: Alex made a conditional agreement that we would wait for our customer on the job site.

What happened later was stored in my memory like an amateur film made by an impatient cameraman and interspersed occasionally with a bunch of sepia-washed photographs capturing the Station's true identity — this is how schematic, discrete and transitory the remembrances of my first jump were.

We approached the Station. A very long joint reactor building immediately struck the eye with its monumental appearance. Next to it, silvery-white masts and an elaborate web-like wire network shone in the bright sunlight. "*ORU*,[16]" said Alex.

We turned right, crossing a narrow channel via a two-lane bridge. Traffic both ways was very intense. The amount of specialty equipment amazed me. There, for the first time I saw a "mixer" with concrete, assembled with a truncated cabin only for the driver. The cabin wrapped, like a cocoon, with multiple lead sheets, with only a few tiny windows — their frames were bolted, holding very thick, polarized (due to the high content of lead) glass. Huge rear-view mirrors were installed everywhere — on the wheel covers, on the door frames — apparently to help the driver to see better when the mixer backed up. The machine looked impressively militant.

Judging by the small bust of Lenin, installed in the middle of a small square tucked in front of the conjoined 4–5 story buildings, we had reached the central square, and the structure in front of us was the Administration Building of the ChNPP. To the left of it extended that impressively lengthy structure, which I noticed from a distance — the main reactor building. It connected all four reactor units in one ensemble, maybe a kilometer long, no less; it was also tall, at maximum height reaching perhaps 50–60 meters. Our *bort* maneuvers between vehicles seemingly chaotically moving about the square. We miraculously avoided a collision with a *KrAZ*[17] whose windows were fully covered by thick metal plates; only a few narrow holes provided visibility for a driver inside. Like the mixer, the *KrAZ* cabin was also heavily armored with lead sheets. The *KrAZ* was backing up at quite a high speed.

---

[16] (Russian: *Otkritye Raspredelitelnye Ustroistva*) Open-air high voltage switchgear.

[17] (Russian: *Kremenchug Avto Zavod*) The heavy three-axle dump truck.

"Blind b*tch!" Vasily hissed, frenetically yanking the shift stick and again hitting my knee hard.

We passed the tight drive-through opening in the Administration Building. A few seconds later, we abruptly stopped. The driver hit the steering wheel in displeasure. Vasily and Alex were disappointed. The dirt road to the left, along units 1 and 2, was barricaded by concrete blocks. They were installed recently, Holodov explained, grimacing; earlier it was possible to scoot by a *bort* almost to *ABK-2*[18] — a center for ChNPP personnel to get prepared for work shifts. It contained an elaborate system of dosimetry control, clean-up/showering facilities and clothing exchange, which since the accident was slated to serve the liquidator teams. We had to change our "clean" clothing into "dirty" (in radioactivity sense) uniforms there. Now the barricade forced us to walk all the way to *ABK-2*, which meant, as Alex explained to me, unnecessary additional exposure.

We unloaded the *bort* in the shade of the main reactor building.

Vasily quickly drove off; the letters *MM 00-02* on the *bort's* back waved goodbye to us, while the truck rolled over dirt piles and puddles.

We hurriedly moved to *ABK-2*.

Sweat ran in streaks from the forehead and stung the eyes. The heart beated strongly, echoing in the ears. The presence of invisible, intangible danger was very disturbing. I knew that it was purely psychological, but recognition of the fact meant nothing. *I was getting exposed to the radiation — for the first time in my life it was not an X-ray, it was far more dangerous ...*

The appreciation that there were dozens of people around me was slightly comforting. They were calm, focused, going about their businesses. Alex was also quiet and absorbed. My initial stress was tampered by "technological noise": the sound of running motors, pumps, metallic clanking, hissing of compressed air, that were coming from different sources around me. It merged into a familiar background that was reassuring for an engineer.

---

[18] (Russian: *Administrativno-Bytovoi Kompleks No. 2*) Administrative-Household Complex 2.

Gorin, our *dozer*, quickly adjusted the telescopic rod of his *depeshka* and took the radiation level readings on the go. During the night briefing, Alex and Igor warned me that the radiation levels, *"fons"*, changed completely unpredictably at the Station, depending on the location, time, wind direction and wind strength. One additional weird thing was the so-called "gamma-beam", a phenomenon rarely observed outside the main reactor building, but "popular" in the technological hallways and rooms nearby the crippled unit 4: for unknown reason, there were some sorts of strong and persistent one-dimensional emission of gamma-rays, similar to laser beams, and one had to run like hell in order to pass those spooky areas, otherwise it meant big trouble.

We briskly walked on the dirt road. Sound of boots loudly hitting the dry ground, heavy breathing of troops elevated the anxiety.

"Thirty milliroentgens... Fifty... One hundred... Half-roentgen... One!" *Dozer* monotonously muttered.

Without waiting for a command, the team started running.

I saw along the way quite a lot of dull grey-colored concrete buildings of different shapes and sizes and, as I figured, of various technological uses. Some of them had glass walls, top to bottom. Sometimes we ran under multiple pipelines, bundles of insulated wires, conveyor corridors on sturdy concrete stilts, which cross the territory of the Station's Industrial Zone in an elaborate pattern — typical and well-known to me as a technological necessity of any big industrial setting. I noticed that in almost all of these small and big buildings there were people —almost none of them was empty.

"*Otsidka.*[19] Inside the buildings, *fons*, the radiation levels, are lower by a minimum of ten to twenty times — no dust, no dirt. Everyone can relax, have a smoke," explained Alex on the fly.

In front of every *otsidka* there was a low-cut steel trough. Most had stationery inlets and outlets of running water. I understood the purpose: it was there to wash off the radioactive dirt from the shoes. But where did the contaminated water go from the trough after the shift?

---

[19] Temporary shelter, rest area.

Progressively, stronger with every second, a pulsating growl of a heavy helicopter took over and subdued the rest of the noises. I couldn't see it, but I was mesmerized by its powerful sound.

While moving, we went around two identical apexes of the smaller reactor units 1 and 2, protruding from the gigantic reactor building — they were painted in darker tones and were easily recognizable from the distance.

On the right, I saw a mammoth-sized concrete chimney stack. The sun came out from its silhouette, and I suddenly realized that it hurts badly — and instantly, bringing a piercing headache — to look not at the sun itself, but just in its general direction, or even at brightly lit objects; my eyes instantly became watery and shut down. This was not the usual protective reaction. It really hurt to open my eyes to look anywhere close to the gleaming disk.

(Later I learned that this was one of several signals of the radiation affecting my body, and this particular reaction — inability to look toward the bright light, especially sunlight — was certainly an attribute of the low radiation levels, around several hundred milliroentgens per hour.)

The running, even at slow pace, wore me down. I felt relief when, after turning around the corner of the "Two", I saw a multi-story white cube of *ABK-2*. We used a backdoor entrance.

"Get changed ASAP. Meeting point — lobby downstairs, fifteen minutes tops!" Holodov commanded the team.

We sprinted up the stairs to the fifth floor, assigned to our Brigade. Alex opened the door and immediately an intense smell of the tired, worn out men's bodies hit me. It was mixed with the rancid smell of disinfectant, cheap soap, and tobacco smoke.

There must be thousands of guys in there!

I was dumbfounded. Holodov patiently dragged me through the maze of lockers, making his way toward the windows. Along the wall was the last row of lockers, with wooden benches at the side. Sound of running water and humid air indicated close proximity to the shower stalls. Amazingly, in this corner there were far fewer people than near the entrance. Alex found two vacant lockers. Opening the door, I recognized the odious *propitka,*[20] a curse of the Soviet Army. This "brilliant" invention of some Stalin-era military research institute seemed to be an indispensable attribute of our Army for ages, intended to protect the troops against all kinds of weapons of mass destruction. During military training, we certainly had our share of endless roasting and melting in these Torquemada-inspired outfits, because by design they did not let *any air* get in, either fresh or contaminated. It could be easily recognized by the oddly sweet smell of the impregnating composition; however, there, in a phenomenally strong mixture of aromas, it was almost indiscernible.

"Get dressed," said Holodov. "Do not forget to use an underwear shirt, otherwise your skin will start itching badly in less than half an hour. Stupid impregnation..."

Amazingly, within minutes the room was almost empty. I believed that there must be a way to organize this havoc better, but that was certainly not my job.

For a few moments, I observed how a naked man, who had just stepped out of the shower, walked toward a line of polyethylene bags near the wall and pulled out a set of pants and jacket, a shirt and a pair of socks.

"Some clothes are washed here, in the Station laundry, where our AAS crew and Igor work, but basically everything else is new," Alex explained, checking his watch. "Hurry... After the shift is over, you dump everything that you used in the bags, over there, at the moment when you cross the door sill." He pointed to several large polyethylene bags with yellow labels at the entrance. "Do not walk to your locker in dirty boots and uniform! Now, leave your boots in the closet, do not contaminate them while working. Over there, in that corner, there is a large wooden chest with boots, they are used many times but should be *clean; dozers* check them regularly. Wear two pairs of socks, take the boots that are one size up from yours, it will be easier on your feet."

He was already dressed. Impregnated uniform, together with the soft cloth helmet that covered his head and wraped around his neck,

---

[20] Impregnated field uniform, designed to prevent the penetration of harmful liquids and vapors.

made him barely recognizable. I tried to get used to the stiffness of the generously impregnated brand-new uniform. It did not seem to work well.

First-floor lobby was bustling with liquidators like a flea market with tourists. Our guys were already waiting for us, gathered at the stairs, away from the windows. It was impossible to recognize any of them in the outfits, especially since I saw their faces only for a few minutes at the Brigade dispatch area. It felt like it was a few hours back. I tried to remember the look of our *dozer* in *propitka*, but then figured that a blue tape on Gorin's *depeshka* was easily recognizable.

Holodov recounted the troops, calling their last names aloud from a small notebook, and led the team out.

We moved at a brisk pace. The sun immediately started frying my back in *propitka*. I also quickly realized that the cloth helmet had a cotton wool insulation; inside it felt as warm as the winter fur hat. But I was not complaining; in my mind, that was exactly the time when it would not hurt to have more of *anything* between the radiation and my body.

We ran under some technological overpass. Using its shade, I looked up. High above, seemingly hanging over the edge of the reactor building, I saw the deaerator stack,[21] whose white and red stripes would later become emblematic for the clean-up and containment operations. I felt nervous. The unforgiving *propitka* faithfully isolated my body from the outside air, and the underwear shirt already clung to the skin like a shroud, floating in the excess sweat. Nervousness changed to feverish excitement; something — some unusual feeling, light tingling in the lung alveoli, like the bubbles of divers' disease — began boiling in the blood. Legs, which only a few minutes ago were strong and obedient, suddenly weakened.

Holodov yanked me by the sleeve.

"What are you staring at, let's go!" He pulled me out into the sunlight, and I squinted like a vampire from a sharp pain in my eyes, which immediately became watery.

---

[21] The chimney of deaerator, an iconic red-and-white striped symbol of Chernobyl accident.

"Whenever you get a chance, find yourself a pair of sunglasses," Holodov continued tutoring me on the way. "The eye retina is more fragile to radiation damage than the soft tissues, at least that's what I was told here by the guys from the Kurchatov Institute. Even plastic could stop beta-particles, let alone glass, so any sunglasses would do."

He suddenly stopped.

"Gorin, not that way! Veer right, along that wall! Come on, people, look sharp and go, go, go!" Screamed Alex. He left me and rushed to reach the team that went ahead, huddling around our *dozer* like chickens around a hen. It was instinctive, I figured, Gorin was the only one who knew the answer, at least some of it: *how much radiation we are absorbing right now...*

I followed them, dragging my feet in oversized boots. I had to give it to Alex — socks were far better than foot-cloths. I also learned that my lungs definitely disliked the quality of air passing through *lepestok*. It had a weird sweet odor, which made me nauseous.

It was a seemingly endless race under the scorching sun across the open area, which appeared to be scraped by an enormous scalpel that mercilessly ripped off the upper layer of the soil. The chaotic jumble of tugged, torn and bent pipelines, hoses, cables, large chunks of broken concrete blocks, spreading rusted steel armature, all mixed with dirt and covered by something like foam, prevented us from running straight, and we dashed in zigzags, as if we were under fire.

We approached a small two-story building. The corridor with large glass panels connected it with another building about 50 meters away. Exhausted, the team crushed inside through a single leaf glass door of the first-floor hall and crowded around the *dozer*, impatiently mumbling. Gorin lowered the probe to the floor.

"Twenty milliroentgens ... It's good, people!" He sounded happy.

Hateful respirators yanked down, shaking hands pulled out cigarettes.

"What was the maximum, while running?" Holodov asked the *dozer*.

"Five roentgens per hour. Closer to the entrance," Gorin replied, trying to light his cigarette with trembling hands. His lighter had metal casing. I wanted to tell him that it wasprobably not a good idea to carry it *there* and then bring it back to Brigade, but I kept quiet. I was sure he knew, and he chose otherwise.

"It's nothing, you guys! Only a half roentgen for each until now, and we have almost reached the "Four"!" Alex clearly wanted to encourage the troops. "Not to worry, we work fast, we work as a team, helping each other, and if we do that, I swear that no one will get more than two roentgens!"

The building seemed unoccupied. We were the only ones there, and the stale air suggested that it was abandoned for a while.

"Are you sure that we have come to the right place?" I asked Alex quietly, and then instantly regretted it. He looked down.

"This building belongs to the ChNPP construction management. The Chief Construction Manager is Kizima Valery Trofimovich, his office is in that wing. Go wander around, it's a free excursion," he smiled sadly. "We have spare time before our customer comes. In Kizima's office, on the floor, there is a pile of his business cards, take one for a souvenir if you want."

He paused for a moment, then continued much louder, talking to the whole team.

"If the customer does not arrive in twenty minutes, we scram back."

The troops exhaled in relief.

I felt awkward. I craved action. *This is my first jump, and now what?*

I walked halfway through the connecting hallway, stopped and lighted a cigarette. My heart resisted calming down. Adrenaline still shook the brain.

Then I noticed something strange. On the windowsill, I saw a dead butterfly. A few centimeters further, there was another one. And then another, and one more, more, and more ...

A few hundred bright-yellow butterflies, suddenly caught by an incomprehensible death, lied there. Did they really die of radiation? They say insects could tolerate high levels of radiation, why then did it happen?

It looked symbolical, almost theatrical, like in the (better) horror movies, when something seemingly neutral brought a new psychological shock, an unexpected twist to the whole story ... yet somehow, I felt that this was not a movie.

This is real.

Inadvertently, I caught a reflection of dead butterflies in the glass. I looked outside through the window.

I will never forget this moment.

All that I saw until then at the Station was intact, even if it was deserted and dirty, but still undamaged. But there ...

A huge building directly in front of me was crushed right in the middle, as if a giant fist furiously pounded and smashed it. There was no roof or walls on both sides of a huge wreck of uneven forms, demolished from top down to about a third of its height, with frightening signs of sheer annihilation. Disjointed concrete blocks were hanging on thin metal rods; their smaller fragments scattered everywhere. There were broken, warped metal structures, mangled pipelines, crumpled equipment, rusted or burnt, all covered with soot or thick layer of dust and ash.

There it was, right there, a few hundred meters away from me, the monster of nuclear calamity, the dragon with the most perilous vigor able to eradicate the whole of mankind ... I stood speechless, feeling that my brain suddenly froze and my legs refused to move — and somehow, I did not want to walk away. I was awestruck by the most intense view I had ever seen in my life. I physically felt a menace that was coming from the shattered reactor.

The magnitude of destruction was more visible closer to the epicenter of the wreck, where the pitch-black ribbons of soot covered most of the remains. On the wall facing me I noticed bundles of ropes, clinging to the walls like giant grey roots. I also saw the ugly greyish "tablecloths", covering the ruins in several places. Could it be lead? It seemed like it was, and it apparently melted due to intense heat coming off the crater. That's how they dumped lead in the crater, I figured: rolls of lead were hanging from helicopter on the ropes, and were dropped down the reactor.

I noticed that multiple hits had missed the target, the center of reactor containment, and wrecked the remaining reactor hall panels and even nearby buildings. The brightly painted deaerator stack on the roof of the building, alongside the exploded reactor pit, appeared ridiculously fancy next to the chaos of destruction.

The total absence of people in the area around "Four" was unnerving, considering how busy the front part of the Station was.

I did not blink for a while, and tears started accumulating in my eyes, corrupting the image, which came across wavy and distorted. I thought that I spent only a few seconds looking with inexplicable captivation at the demolished unit. I was bolted to the floor, although realizing that it was probably not a good idea to stand in direct sight of such a fierce source of radioactivity. I was unable to shake off that animalistic primeval fear that had lived in every human for thousands of years — the one that raises the hair on the back of our necks when we see something galactically dangerous.

And then I saw him.

A welder worked at a close distance — a few hundred meters, give or take — right in front of the destroyed reactor, on an absolutely open, unprotected space. He was slicing a warped mound of metal into pieces using an acetylene torch. Sparks were splashing down and to the sides. As far as I was able to see, he had some sort of protective gear, because he looked a bit bulky, and his face was covered with a heavy two-can respirator and goggles. He worked methodically, professionally, periodically moving the torch aside and checking the quality of work. I was sure that he was even whistling.

No too far from him, the monstrous dragon of unchained nuclear energy was relentlessly exhaling vicious radiation, but he worked focused and composed.

And totally unexpectedly, looking at him, I also calmed down. For the first time that day, I foiled my fear.

I picked up Kizima's business card in his office. Just like in Pripyat, everything in it was screaming about a momentary, dramatic evacuation. Everything was frozen in time. Maybe in the morning of April 26th, or later that day, when the owner of the office was already on his way to the hospital (I learned later that he was severely overexposed, because he tried — as many of the middle level managers of ChNPP — to help any way he could, but *anyone* who was in the close proximity to "Four" those first hours and days, was receiving a harsh and hasty reality check, with graphite, fuel, and *TVELs*[22] covering the grounds all over the premises. So, they were told to scat immediately, to drop everything they were doing and leave. There was an open folder with some papers on the table, stacks of documents, blueprints, drawings, a switchboard with a bunch of telephones, lamps, furniture … everything was abandoned. For a long time, if not forever.

That business card became my lucky charm for the next forty days of my Chernobyl.

The customer did not come. We returned to *ABK-2*, let two exhausted *dozers* to checkour clothes and boots at the entrance, then dropped "dirty" clothing in the yellow bags, showered, and waited for a while for Vasily and his frightful "MM", "*Mogilnik* Dream". When he came, our anger quickly evaporated once we found out that he managed to snatch from the ChNPP canteen, where only Station personnel were eating, a 20-bottle case of sparkling water, which we gladly emptied in no time. Radiation makes you parched in the extreme, it desiccates you like a food dryer; you want to drink non-stop, like during those distant children's days of summer football, when after the game you craved to quench your thirst, to drink and drink tooth-freezing cold water from the tap in the school basement.

To my downright fury, my brand-new boots were stolen from the locker. Alex was sympathetic, but since the cabinets had no locks on them, that happened quite frequently.

The problem was corrected by the *dozer* at the level 5 entrance. Pitying the newcomer, he winked at me, went to the locker where he apparently kept his things, and with a gesture of a magician pulled a pair of new high-cut officer shoes — genuine leather, laces, fancy welt, heels, all the works. "Use them for a good cause, I can get myself more!" he said with heavy but warm Western-Ukrainian accent.

The shoes looked chic, right size and all, but did not mesh well with my breeches, showing the world the gleaming white socks in the wide gap between breeches and shoes. Alex chuckled and commented that I looked like a POW; therefore, on our return to Brigade, kind Vasily gave me a pair of slightly worn khaki trousers, part of officer's light field

---

[22] (Russian: *Teplo-Vydeliayuschie ELementy*) Heat-generating elements.

uniform, designed to be worn with shoes. I used them throughout my Chernobyl stint, as well as the shoes gifted to me by the *dozer*.

Experience teaches the teachable. Since that time, I always left the shoes in the *bort* cabin, carrying with me "interim" shoes to get to *ABK-2*. This was well-managed by white canvas boots, used by the Station personnel. In these white shoes and in the clothing combining two different types of field uniforms, I certainly looked ridiculous, but I did not care. At least, such a complicated scheme of changing the shoes guaranteed much better chance of not dragging extra roentgens with dirt to Brigade and to my tent.

On the way back, the *Moldovan* checkpoint personnel who were usually outright unforgiving, by definition of Vasily, unexpectedly relented and let us out of the zone without sending back to *PuSO* — a chubby, super-tired Sergeant-*dozer* shook his head and said that a sacred **MM 00-02** must be already *glowing* at night, that's how much secondary radiation it absorbed.

Our commanders did not scold Alex for a missed work day. Moreover, Zelenov suggested to him to squeal to the *SGC*,[23] explaining how our team was exposed to high levels while idling and waiting for a civilian customer who did not show up at all. This could have meant a substantial punishment for a guy who let us down, so I had no doubt that Alex did not write anything.

The next morning, very early, while in my dreams I was gazing down at the edge of the crater, while the welder was sealing the "Four" using a huge cutting torch, Senior Lt. Alex Holodov departed from the premises of 25th Brigade of Chemical Defense.

When I woke up, I found a carton of cigarettes ("Chesterfield"!!!) that he left on my night stand, and a piece of scrap paper over it that said:

"DON'T BE A HERO!"

---

[23] (Russian: *Pravitelstvennaya Comissiya*) Special Government Committee, the leading government organization created immediately after the accident to deal with all aspects of the accident aftermath.

August 5, 1986
Oranoe village, Camp of the 25th Brigade of Chemical Defense

"Delegation of the Central Committee of the Communist Party of Ukraine has arrived this morning to deliver the high award to the Brigade. Warriors-chemists solemnly swore that they would honorably bear the red banner, awarded by the Central Committee, continuing their relentless daily work to make a significant contribution to the elimination of the consequences of the Chernobyl disaster. A ceremonial march was held, when the columns of Battalion of Chemical Reconnaissance, three Battalions of Special Operations and other units of the Brigade went by the podium with distinguished guests. Many troops went to fulfill their duties at the Chernobyl nuclear power plant immediately after the parade ... "

(From the newspaper "For Our Soviet Motherland")

"Good morning, Sir," I grandly bow my head.

"I trust you slept well, Sir?" Replies Svyatoslav, a.k.a. Svyat, gallantly tying a thin towel around his neck as a scarf. He wore navy-blue ChNPP technician trousers, operator's white boots (their fabric was cut to the sole on the back to make slippers, his own proud invention), and camouflaged reefing jacket, which was thrown on bare torso. Svyat was taller than me, skinny and big-headed. His eyes behind thick tinted prescription glasses were red, swollen, and watery. He suffered from an exacerbated stomach ulcer and insomnia. Last night we watched "The Mystery of the Blackbirds", Soviet movie, adaptation of Christie's "Pocketful of Rye". In the Brigade's improv cinema (several dozens of rough wooden benches with no back support and white linen screen stretched between two poles) we watched all the best movies of the last decades. British high life in the Soviet movie interpretation was weird, but who were we to judge?

"Indeed, I did, Sir. I am, however, outraged. My butler, such a *quisby*, again forgot to walk the dogs," I shook my head in a fake sorrow.

We slowly walked in orderly manner to the washbasin.

A person who is unfamiliar with the Army life would be awestruck by the abundance of green color of all shades in the field camps: tents, machinery and equipment, uniforms on military personnel, together with grass, shrubbery, with the forest as a backdrop. The camp was full of life and energy during the early morning hours, but we were relaxed and sleepy. Svyat had worked two shifts in a row and had a break that day, and I had served as a Battalion Officer-on-duty formally until 24:00 hours last night, so I was resting as well.

The washbasin was meant to serve about three dozen people at once. Typical Army camp setup: a large cistern from *ARS* was set on high scaffolds, slightly tilted to the front, where hoses connected it with two long pipes, sixteen tabs (or using the Army language, "nipples") in each. Pipes were hanging over long wooden pigsty-type basins, again posted on scaffolds at waist height — simple and efficient. Rainbow-shiny puddles under the sinks emitted a mixed odor of toilet soap and toothpaste. The smell was deeply rooted in our memory since the glory days of summer pioneer camps. I felt like soon there would be a hot breakfast with mandatory oatmeal, then the whole camp would line up and report to the Senior Pioneer Leader ... I blinked rapidly: it's either a heart-warming memory of my pioneer past, or the bleach that, in the fear of infections, was quite generously poured into any body of water, from the hundred-liter tea and soup pots in canteen to the cistern.

I shaved, regretfully examining liquid-filled sunburn scabs on my ears in a shard of mirror affixed to the scaffold. The neck surely did not look better. There was no way to shade the open areas of the head and neck from the pitiless sun because my hat had no protective flaps, and sunblock was the luxury of a civilian past.

The fifth day of *my Chernobyl* had officially started. Wait, I couldn't call it "a day in Chernobyl". It's interesting that the name of a town that was farther away from ChNPP than Pripyat, gave the name to the Station, not Pripyat. Besides, it was already my fourth torturous day in the Brigade camp, not at the Station.

Thoughts about civilian life, about the institute, about work, about my research projects (*what projects?!*), caused nothing but a dull perplexity. A tiny photo of Lora rested deep in my wallet; too many memories were linked to it, and I did not want to get distracted at the moment. The Army's relentless capability, if not the art, of suppressing individualism swallowed me rapidly and fully. I had already educed the secret skill of high-speed attachment of a fresh white under-collar, as well as intuitive skedaddling from commanders and *saluting in motion* (this expression in Russian has much deeper, funnier meaning since it almost literally connotes losing most cherished possession of every girl while moving). I easily recalled the cliché Army jokes. I masterfully walked in formation with other reservists. All in all, as many of my fellow liquidators, I quickly converted my civilian self into some sort of pseudo-Swiss Army knife capable of imitating all mandatory functions of a military man. I used the word "imitating", since none of these functions were in fact helping my main purpose: *clean up the nuclear mess.*

(I want to pause for a short while and pay tribute to my fellow liquidators, Army reservists. Every one of us then, in 1986, had an eerie, hopeless sensation of an edgeless gap between the job that had to be done at ChNPP and the cartoonish, grotesque life back in the Brigade camp, where our adored Army tried to instill the schedule, the rules, and the conduits of the *real* reservist duties on top of our dangerous stressful work at the Station. Sure, under other circumstances, during a regularly called reservist duty, we all — with various degrees of engagement and enthusiasm — would have taken part in the mandatory marching formations, sung patriotic songs, had the daily duties of raking the sand around the tents, picking up trash, painting the fence, and read aloud the Party newspaper during "*Political information*" gatherings. The irony of the pure existence of that gap, which, in my understanding, had never appeared in the minds of our commanders, was turning our lives into a *commedia dell'arte* when we were back at our tents — tired, sick, sweaty, unable to respond and to act the way the Army reservists were supposed to. We lived through this mockery when we were supposed to rest, and continued to do our job back at the Station — over and over, again and again.)

... I carefully lathered my cheeks up a second time.

Svyat was losing patience.

"I wanna eat, man. Today is the joy of a fat boy, the long-promised processed cheese," my companion pushed me.

"Yep," I started shaving again. "I hope they taste just as good, as they smell bad. Save a couple for me, I will carry them under my heart until they ripen."

Processed cheese. The breakfast of champions. I had nowhere to hurry.

For some reason, the commanders kept me idle. They did not send me to the Station, but twice in four days made me an Officer-on-duty in the battalion. Atrociously bone-headed mission, as we unanimously decided with Svyat and Lieutenant Sergei Petrov, Deputy *COS* for Captain Zelenov. We talked briefly the night before, after watching a movie, while smoking outside the Command Tent, Sergei's fiefdom.

(Petrov was a short, chubby guy with a face resembling a cupid — blue-grey eyes on a roundish cheeky face framed with curls of blonde hair. On that, the similarity with cherubim ended. Sergei was a cold-blooded computer on two feet, precise, inquisitive, smart; his endless stamina, efficiency and ability to multitask had earned respect of both upper commanders and peers. Svyat and I loved his cynicism and lack of emotions, which were painted over with nearly a childish feel of camaraderie. The three of us were tight during those sun-drenched days in Chernobyl.)

"Generally," Petrov said, "we cannot use the personnel from our reconnaissance platoons, either squad commanders, or drivers, for the Station jumps. Replacements did not arrive for over a week, and several old champs had exhausted their quotas of twenty-five roentgen."

He knew what he was talking about. He and Zelenov controled the Battalion personnel records — all data related to our doses, daily and cumulative, data of radiation fields related to Battalion reconn operations, daily jump planning, replacement requests, sick bay listing, salary of staff on active Army enlistment, personal information, you name it.

"Remaining several officers, who are not tied up with daily reconn, and still under twenty-five, like you and Igor, are extremely scarce, and besides, Igor is strapped with his miners to the Station laundry," Sergei continued. "So, I am saying: do not sweat it, your time will come again, and very soon!"

I fully understood and agreed, but so far it was day five, and I was stuck there, "on premises", with no way of telling *when* I would get called to jump. I just couldn't stand wasting time while hanging around in the Brigade. The first jump was then pictured in my head as something nearly Hollywood-like, which happened to someone else. The stress, the risk, the adrenaline that were so vividly felt at the Station, were indiscernible there, during four days of idling. However, much more real were the endless preparations for the Government Delegation's arrival, scheduled the week after: full Brigade practice (marching in formation every morning with singing, mind you!), washing of anything that was washable, painting of anything that was paintable. Brigade commanders were freaking out, and this was telegraphed down the command chain loud and clear. For a bystander, life in the Brigade camp was full of meaning and purpose, only the nature of that hustle was clearly alluvial, and plainly had nothing in common with the colossal event that was developing at the Station. The commander's logic was unidimensional: the soldier must be busy, otherwise he would have extra time to think, and thinking was not his major function in the Army!

The day before, at morning practice, the heat at 08:00 was already unbearable, yet the brigade was sent marching three times, going by the improvised "podium of esteemed guests". Preparation for the passage of a solemn march in front of the bright eyes of the delegation. Guys marched without the required enthusiasm, I figured, hence we were ordered to go three more times — and apparently, we could've marched more, if not for some poor fellow from the First Special Operations Battalion, who conveniently fainted because of the immense heat and fell face down, slitting his chin on the asphalt.

The story that I am going to tell now perhaps is not the most interesting, but very funny and — again — underlines the weirdness of our dual status as liquidators and as Army reservists.

The Reconnaissance Battalion quarters and the Brigade Command tents were separated by the asphalt-covered road, leading off the highway into Brigade premises. On our side, the sleeping tents were placed next to the road, followed by general purpose tents, then soldier's canteen, officer's food block and on the far back one big latrine for soldiers and four individual toilets for officers. Since the Brigade Command tents did not have their own toilets, there was a steady ant chain-like traffic of high ranks to those four wooden "need-houses" through our Battalion quarters all-day long. The frequent chance to meet face-to-face with all the bosses from across the road eventually led to the very confrontational but, sadly, brutally enforced result for my fellows in Reconnaissance Battalion. I am referring to the fostered among liquidators-reservists disregard to wear the Army statute-required field uniform. It actually happened inadvertently. As I already mentioned, the turnaround of clothes, shoes and personal protective gear at and around the Station was out of control, both in terms of the safety precautions, and in terms of the lack of spare uniforms of the same style. It was plainly impossible to provide spares for thousands and thousands of active Army troops and reservists, particularly since the disposal of contaminated uniforms occurred on daily basis. Of course, laundering was employed as a method of uniform deactivation, but again, rotation and replacement of uniform was not well-maintained at all, therefore many of the liquidators used any spare *clean* (in radiation sense) clothing available, and at the Station that was almost exclusively civilian. This was the reason for some absolutely unconventional combinations of uniforms that were used in our Battalion.

One day, when some high-ranking commander was rushing to the sacred four-unit "need-houses", he caught a guy who, according to our BC^{24} Lt. Colonel Golubev, was wearing only "a ridiculous T-shirt, officer's trousers and a pair of sneakers", and who was coming out of one of the officer's toilets with a newspaper (apparently, the fella enjoyed reading the news while using the facility). Seeing the Colonel who was impatiently waiting outside, the violator disappeared so quickly that the angry commander was unable to remember his face.

Golubev, apparently, went through an intense reprimanding session with the outraged superior officer, because after dinner he called for an urgent officer meeting. The case was painted by enraged Golubev in

---

[24] Battalion Commander.

much more tragic colors than it appeared to us. However, the outcome of this comic story was obvious to all, and our BC ended the meeting with stern verdict.

"If I ever catch any one of you or your *guerillas* outside the tents not dressed in the statutory clothing, I will personally see this idiot court-marshalled!" He rumbled. "What is this — an Army or a brothel? What's next, I ask? Hookers on every corner?!"

We quickly looked at each other.

"That'd be nice," Someone behind me whispered.

The glorious day of the Government Delegation visit was approaching with the inevitability of a typhoon. The delegation had to award our Brigade an honorary banner of the Central Committee of the Communist Party for "substantial and continuous impact in the liquidation of consequences of the accident at ChNPP" — the obscure phrase that drifted from Gorbachev's televised appeal a few months earlier and became a cliché. As always in such situations, our Brigade commanders were seriously flipping their lids, and tried to calm themselves in every way possible, handing out to us even more bizarre tasks. This madhouse, obviously, was first and foremost affecting those poor souls who were not immediately engaged in various activities related to the reconnaissance or clean-up.

For example, me.

After breakfast, I was assigned by BC Golubev to work, together with two soldiers, on covering *"perimeter of the tents"* (as stylishly expressed by my commander) with turf slabs. I actually found a certain logic in this task, as dust and sand were constantly hovering over the territory of the brigade; radioactivity wasn't high as compared to the Station levels, but dangerous enough to increase the *fon*. I, however, had no illusions and was sure that the top dogs, who gave this order, were certainly guided by considerations of aesthetics, not radiation protection.

While my two guys were carrying the turf on stretchers with the speed of zombies, I was cutting a thick turf edge of the forest glade with a shovel, and at the same time exercised my brain with simple math. The result was totally maddening. The generally accepted average level of radiation on Brigade territory was 30 mR/h. My total share of absorbed radioactivity had to be 25 R. If I substract 1.75 R, which was written as my cut for the first jump, from 25 R, and then divide the remaining dose by 30 mR, I should spend *about 775 days* cutting turf to fill the quota!

Devastated by this math, I took off my jacket and kept working in the T-shirt. Screw BC's warnings about the uniform.

Even in the shade of the trees, the heat did not get less intense. Svyat swore that during the previous three weeks not a drop of rain had fallen in the 30-kilometer zone, because the State Hydrometeorological Agency, *Gosgidromet*, was constantly spraying a solution of silver iodide from the planes, high up in the air, preventing the formation of rain clouds. It sounded very cool, but I still doubted the efficiency of such a measure. Torrential rains were common there in the fall; they could lead to a sharp rise of radionuclide concentration in Pripyat aquifer, washing the particles through a sandy soil. As a consequence, nuclides could break into Dnieper river, and then into the Black Sea.

I was desperate and bored; I physically felt the lousy flow of time. By noon, we must have already covered with turf not only "tent perimeters", but the whole Battalion grounds. At some point, my "zombies" brought a bottle of sparkling mineral water, which I drank straight up. The warm, salty "Mirgorodskaya" did not quench my thirst.

Lunch was not far away. I reminded myself of a prisoner, because I measured the time by food consumption — "from breakfast til lunch". It was really sickening.

Sensing my mood, the soldiers also slowed down. I began to wonder if zombies have the ability to vary their speed. We took lunches with us from the canteen to the glade and ate as thoroughly and as slowly as one could eat a split pea puree. We continued after lunch at the same sluggish pace. Around 17:00, we were told to stop. I was unable to express any joy, and so were my reports. I let them go and went to check out the mobile shop.

The arrival of a mobile shop was one of the few entertaining moments in the otherwise dreary life of liquidators in the Brigade.

The merchandise assortment, which was usually delivered to us those days, had no common sense whatsoever and deserves a special mention. I figured that the selection of stuff to be sold to liquidators was following the old-style trade principles, laid out centuries ago by the Europeans dealing with Navaho or Aleuts. Perhaps, some high-ranking supply chief took pity on us and ordered for deliveries, among truly needed and requested things (like cigarettes, shaving cream and razors, soap, detergent, candies, chocolate, spare batteries and so on), some oddities like Turkish style sweets, various juices, Chinese thermoses and flash lights (those were in high demand, because very often the Brigade generator was cut off after midnight, and — as on the night of my arrival — Brigade premises became pitch-black). What was especially gratifying was that the quality of all the merchandise was exceptional and for the most part consisted of imported goods: cigarettes were mostly Bulgarian (those days, the Soviets enjoyed Bulgarian tobacco quite a lot) and even American — "Chesterfield", which Alex left for me when he departed. These made a rare appearance but were delivered a few times. Pepsi Cola, a rare treat for the Soviets in the 80-s, was brought in frequently, although in limited quantities, and thus was quickly gone in a matter of minutes.

I spent some time chatting with the salesperson, a well-mannered middle-aged Senior Sergeant who remembered me from this talk, and always kept on the back a couple of Pepsi for me. Finally, I bought spare batteries for my flashlight and went for dinner.

Along the way to the officers' food block, I looked with slight sadness on AAS station behind barbed wire of the car park; it looked completely abandoned, resting on flattened tires. It was naive of me to assume that someone would use the principles and methodology for calculating the doses of a direct nuclear strike to predict the radiation fields formed by radioactive fallout after the "Four" explosion. Spots and bands of radiation fields were the two very typical features inherent for the fallout. "*As leopard skin*", they used to say at the Station. That's why our *BRDM*s were raiding the territory of 30 km Exclusion Zone on a daily basis; it was paramount to know the trends in ever-changing radiation levels within the Zone.

Dinner was mostly uneventful and was somewhat brightened with a discussion — do strongly emitting pieces of *TVELs* on the roof of the "Three" glow at night or not? Opinions were divided; however, the absence of meat on the dinner menu took the heat off the conversation. Svyat, arriving from the route trip, said that at Yanovo, the railway station, somewhere on the back tracks there was an ordinary-looking freight car. On its wall there was a spot, about 2–3 square meters, which had been constantly *shining* 50 R/h since the blast and the first wave of radioactive fallout. He said that it was so steady that he could calibrate dosimeters at that spot.

After dinner, Svyat, Petrov and I had time to smoke a cigarette before the evening operative meeting, in which our BC casually informed me that the next day I would go to jump the *"Pikalov object"* with a team of ten.

It sounded mysterious and menacing. Petrov told me the story of the Pikalov object: mounds — tons and tons — of highly radioactive dirt were bulldozed by General Colonel Pikalov's soldiers using scrapers and other heavy machinery in early May, rolling the soil closer to the walls of units 3 and 4. It's time to move it away from the Industrial Zone, bury it in *mogilniks* to avoid dissipation of nuclides by soil erosion.

Strangely, this news did not excite me as much as I perhaps should have anticipated. The hype of the first jump had already burnt out.

I fell asleep while admiring the powerful, in-tune, snoring of the miners, to whom I had yet to get introduced properly (I was listed as their second-in-command officer!); they were usually leaving the Brigade very early in the morning, and arriving already in the dark, often after dinner. Judging by the snoring loudness, every one of them had the lungs of accordion size.

August 6, 1986
Chernobyl NPP, second hodka

> *"...Clean-up works were carried out around
> the clock by shifts totaling up to 6,000
> people per day... To reduce the level of
> radioactive contamination of the soil,
> nuclear fuel fragments in ChNPP industrial
> zone were collected and eliminated... in
> addition, the top layer of contaminated soil
> was removed, followed by the filling of the
> territory with gravel and its cementing. The
> thickness of such coating was in some places
> up to 8 m. The territory adjoining unit 4 was
> covered with rocks, sand, dry concrete mix;
> additionally, bulk formwork-made cement
> blocks were used... As of August 10, 1986,
> 25 thousand cubic meters of soil was
> removed, and the territory of 187 thousand
> square meters was covered with reinforced
> concrete slabs."*
> (Chernobyl Catastrophe — Part I.
> Historiography of Events:
> Socio-Economical and Ecological
> Consequences)

Tall, thick window panels in vestibule were surrounded by formidable
grey walls of "Three", unit 3 of ChNPP, which were leaning over a
small separator station, where we rested. Our *otsidka* was tucked in
the corner between the main reactor building and the bulging block of
"Three". The space between the *otsidka* and the reactor building was
filled with dirt — in some places, soil mounds were over 10 meters high.
Dirt, heaped there by bulldozers, was interspersed with chunks of broken concrete blocks, pierced with steel armature, crushed bricks, wall
panels, mangled rusted metal pieces and other debris. During the first
days of the aftermath, in the hype of improving the radiation conditions
by all means possible, the top soil layer from the areas surrounding

"Four", which contained pieces of destroyed containment and enormously radioactive projectiles from the exploded reactor, was scraped and literally shoved to the conjoined building of "Three" and "Four".

The way I saw it, it was a good solution, albeit temporary, because next came the realization that these mounds were actually drying out, and it was just a matter of time when the radionuclides buried in them would again begin to spread around by natural erosion — either in the form of dust or in the form of mud, migrating tainted rain water underground.

That's why we were there. That was the *Pikalov Object.*

The distance between our *otsidka* and *IMR* was only a hundred meters. A huge building overshadowed the scene, which should help us to get things done with less sweat. *IMR*'s engine was shut down, the driver hatch was left open. The unit was empty; its portholes, sealed with blue leaded glass, resembled the eyes of a weird, dangerous animal. Track chassis instead of legs; backhoe that was mounted in place of a removed turret looked like a giant yellow scorpion tail which was in bizarre contrast to grey-black patches of lead sheets and blocks, fully covering the armor. They overlapped in many places, which made the machine look muddled, almost creepy.

Next to *IMR*, there were about a dozen large metal containers, with lids on hinges, similar to typical trash cans, only bigger. Containers were scattered around in disarray — they were dropped from *KamAZ*[25] dump truck, which left immediately after dumping containers on the ground.

We had been there for about an hour. Our *otsidka* was one of the subsidiary technological shops that were plentiful on the grounds of the ChNPP Industrial Zone. Captain-sapper, who brought us to the Station from *razvod* and who was responsible for the *IMR* drivers and the current mission overall, explained the function of my team, which was divided into five basic tasks for each soldier:

1. Run from *otsidka* to the point of loading, set the container upright, flip the lid open, then run back to *otsidka*. 2. Wait for *IMR* to fill the container with soil and debris with backhoe. 3. Return to

---

[25] (Russian: *Kamskiy Avto Zavod*) Kama Automobile Plant-made heavy three-axle truck.

container, close the lid and secure it with a piece of wire, so that it does not fold back during loading and transportation. 4. If the soil prevents the container from being closed, remove the soil excess with a shovel. 5. When all containers are filled, the sappers will bring in the mobile crane and the dump truck to load them on. At that time, it will be necessary to provide the usual slinger work: hook the container with slings, assist in the lift-off, help while lowering container onto the dump truck.

Then the dump truck would transport containers over to *mogilnik* — not our concern.

The Captain said all this — including the part of *"not our concern"* — with a soft, quiet voice, addressing the tasks as "simple".

Of course, simple. Provided that you work with *normal* rubble and soil, not with highly radioactive waste.

Before the start, my *dozer* went through the heaps, measured the levels on six points of a rectangle, covering the area of work: four in the corners, two in the middle, all on the run, climbing up and rolling down the mounds. The average radiation level was closer to 15 R/h. The allowed daily dose for work at the Station those days was 2 R per day. The math was simple: we could work there no longer than 8 minutes each. I adjusted this number accounting for additional time that we would spend at the Station — moving back to *ABK-2*, showering, departure — and rounded it all down to 5 minutes.

My team was *eleven people long*, including me, so the total time allotted for us was about an hour; it might look weird, but the task longevity during the jumps was measured by the number of troops in the team. Let's roll.

The first driver ran from *otsidka* to *IMR*, dove inside through the open hatch and slammed the lid shut. The exhausts coughed up a thick plume of smoke, and a loud growling of the working engine reached our ears behind the glass. I sent the first soldier, clocking his five minutes. He quickly straightened the container, opened the lid, and waived to driver — load it up!

I was timing all operations. Task one, preparation of container, took only forty seconds. The soldier returned, gleaming with excitement. It's amazing how much sweat any person produces under the fear of the unknown. His *propitka* already had sweaty spots around the armpits, on his back and chest. Nevertheless, he seemed to be relieved, pulling down the *lepestok*.

"My grandpa can do this while napping!" He said, smiling ear to ear.

This immediately released the tension among others.

The soil was a mixture of reddish clay and sand. Construction debris prevented *IMR* from hauling faster, its backhoe had problem with large chunks, often dropping them. However, the sapper operating the *IMR* was really good, working at moments masterly. He used the bucket to press down the debris, but it made it worse for us, because sometimes garbage stuck out of the container, preventing the lid from closing fast. I was concerned that my guys were pulling the sticks and wires from the container by hands, which were protected only with regular construction cloth gloves, pulling it back, throwing it aside. These oversized pieces would later be removed by loading straight onto the dump truck, explained the Captain.

When all the containers, about a dozen, were filled, sappers quickly brought a mobile crane and a dump truck, *KamAZ*. Only then the *IMR* driver was changed, since inside the machine, protected with armor and additional lead layers, the radiation was not as strong as in the open air.

Then my soldiers needed to tether the containers one by one, so that the crane operator could load them onto the dump truck. Once the load was up on the truck, the soldier had to climb up on it to unhook the ropes from the container. Again, it sounded primitive enough; however, radiation was substantial, it hit your nerves big time, and I was impressed with the focus, precision and speed of my guys out there.

Time went by fast. I went out there as well and also felt the excitement from the outcome. We had already loaded two trucks, and the team sensed the end of the shift nearing. The *spent* soldiers were gathered at the back, relaxing. I was a little worried that our substitution team was not there yet. My next-to-the-last soldier went on loading, which set the timer at about ten minutes before the end of our shift, but as far as I could see, no one was rushing from *ABK-2* in the direction of our *otsidka*.

Observing the work of the team, I thought that the fear of radiation under our circumstances was caused not only by the extreme possibility of getting exposed to radiation, but mostly by uncertainty: *where, how much, for how long?* A nuclear physicist, for example, knows that a piece of radioactive uranium poses a significant threat to his health, but at the same time he is well aware of the limits of this threat, as well as how to defend himself against it. My soldier was filling in a container with radioactive debris, which was emitting 10–15 R/h, and at some point, picked up from the ground (with an almost bare hand) a piece of *TVEL*; incidentally, he knew nothing about it and could hold it for several seconds, completely unaware that his life and death were balanced by this piece. At the same time, he was trembling in fear, pulling some harmless piece of iron out of the rubbish, believing that it was massively radioactive ...

I lighted up a "Chesterfield". Strange sensation. The aftertaste of a cigarette smoked on elevated radiation levels was different. The saliva had a very faint but noticeable metallic taste. It was distinctively different if I smoked one from the same pack in the Brigade. Probably, the reason for this was the presence of large doses of adrenaline in the system while you were out on the high fields. I was unsure, but that was the best explanation I came up with, when mentioning this finding to Svyat and Petrov. In addition, I was slowly getting used to the slight, barely noticeable tickling in the lungs at inhaling the air; it started after a few minutes in the lower fields of under 1 R/h. It felt like someone was gently brushing the lung alveoli with a bird's feather. I noticed it before, during the X-ray imaging of lungs in the polyclinic. During my first jump I already learned that the small fields (single digits R/h) greatly exacerbated the sensitivity of the eyes to bright light. I talked about this with the guys from my battalion at dinner the other night, none of them experienced anything like this. But I do not have to lie; it is what it is.

A walking dosimeter, joked Svyat. I was flattered, and worried at the same time.

The idled *IMR* looked like a toy next to the gigantic windowless walls of the "Three". Dimensions of the soil heaps around the dual reactor building were overwhelming; I figured they rose to over 10 m up near the walls. It looked like an angry giant had thrusted the building into the ground, squeezing out the soil, like paste from an enormous tube.

High above the "Four", a MI-26 chopper was spraying the destroyed reactor with deactivating solution from a huge tank at the end of a cable. The howling sound of the eight powerful blades pounding through the air was audible even through the glass of the *otsidka*. Spraying was done daily, sometimes several times during the day and night, but I had heard that it had no sustainable effect. The condition of the radiation fields around "Four", especially in the immediate vicinity of it, varied greatly, at times quite unpredictably, even within one day. The burning reactor "puffed" with microexplosions, several times per day, because the nuclear reaction in the ruined core was still going on and would continue for many more years. A week before, somebody at *SGC* came up with a suggestion to use a siren at the Station, if the average levels on the site were jumping up, the siren would sound after a new microexplosion. This was considered useful for the guys working on the roof of "Three". However, soon the idea was abandoned: there was no need to seed the panic, people should not be distracted from work.

By the way, earlier that day, on my way to the Station, I saw for the first time how the imported trencher worked — Casagrande, Italian-made, I was told by the well-informed driver of our *bort*. A milling cutter, installed at the bottom of a heavy metal frame, could cut a trench of about half a meter wide to the depth of over 30 m. US-605 raced to install the so-called "wall-in-the-ground", separating the Station from Pripyat River — until then, the driver said with full confidence, not a drop of rain would fall on the Station premises, to avoid the seepage of nuclides to the underground waters during the inevitably approaching fall showers. This was done by dispersing the silver, he said with authority, from the planes all around the Exclusion Zone (if anything, this should have been silver iodide, as Svyat said a while back, but I still shook my head in disbelief).

... My last "commando" returned from the loading area. Time to wrap things up, but the replacement team hadn't arrived yet, and I took a decision to send my guys to *ABK-2*, setting the "jump-back" time (to leave for the Brigade) an hour later. Meeting point — the Administration building hall. Typically, after the shift was finished, the *dozers* had to hand over the report and the sketch of radiation levels at the workplace to the *OGMO,*[26] located in the bunker of the Administration building basement. That day I wanted to do it myself. Weird rumors were spread about the bunker, so I meant to check it out.

I informed the Captain-sapper that my team was *spent* and it's time to get out. He shrugged his shoulders.

"I do not care, and will not wait any longer either, I hate burning my drivers for nothing. We are ready to fold."

However, to his obvious regret, at the last moment, like in a movie, a sweat-drenched Lieutenant almost ripped the door off the hinges, dashing inside the *otsidka*. He somewhat hysterically reported to the sapper that his team had arrived for work at *Pikalov Object*; they were late because their *bort* broke down halfway to the Station. I watched as the Captain's face became red, and I heard annoyed chatter of his drivers. They obviously were not happy to keep going, but there's nothing they could do. The work must continue.

Nice hot shower, ended with quick cold flush, brought me a moment of relief, but as soon as I was out of *ABK-2*, the ruthless sun melted and fried my poor mortal body through the jacket. The way back to the Administration building was more recognizable, marked by the main structures: the cyclopean concrete chimney and the pop-outs of the "One" and the "Two". I am not saying that I was used to the fear or stress, but the unpleasant sensation of psychological pressure, which I had experienced during the first *hodka*, was much weaker.

I spotted some unusual details on the otherwise typical industrial landscape. The abundance of trash and debris was staggering; I was particularly amazed by a stack of small wooden crates that were normally used for tomato harvesting on Ukrainian collective farms. Several trucks and cars, among them one ambulance and a couple of fire trucks, were dumped and scattered along the path from *ABK-2* to the Administration building; they were stripped of tires, the hoods were left open. Clearly, the "car gravediggers" took all the valuables — carburetors, batteries, radiators — with all the radiation that was accumulated in those parts (I learned later that, quite amazingly, it wasn't a show-stopper for the thieves to snatch the valuable parts from those

---

[26] (Russian: *Operativnaya Gruppa Ministerstva Oborony*) Task Force Group of the Ministry of Defense.

cars left in the "Four" proximity, because they were using the stolen parts for the trucks and cars that were kept in the Industrial Zone; quite a clever, yet dangerous, approach). Not a single person was around, and I started feeling creeped out.

No man's land.

Closer to the Administration building the grounds were more civilized. The soil was coated with concrete slabs, which looked more like plates, and they covered almost the entire area of the Industrial Zone behind the Administration building.

The human traffic was also denser. I met a man, an obvious geek, most definitely from the Kurchatov Institute. He had a huge beard, almost the size of a shovel, with a *lepestok* respirator comically covering about thirty percent of that Santa-like hair, barely protecting his nose and mouth. I figured that his beard, apparently, served as an additional filter. The geek, however, appeared to be in a good mood, since he quietly murmured a popular song through the eclectic facial protection. On approaching, he greeted me, gallantly sliding the respirator down and exposing his thick, black beard, covered with white filaments of *lepestok* filter material. I wasn't sure if he exposed himself to radiation because he really wanted to say "Hi" to me, or because he wanted to have a smoke — he lighted up a cigarette on the move. Yep, that's right, he was from the Kurchatov Institute, as I saw on the plastic badge. I was not surprised. Those folks were surrounded by enigma of mystery and almost foolish bravery. They were hugging a devil in their sleep.

I passed under the familiar arch separating the Administration and the main reactor buildings. Central square looked extremely busy; cars and trucks abound, including a couple of *BRDM*. I checked the time: about half an hour, that's all I'd got. I didn't want to keep my troops idle, waiting for me.

The bunker entrance looked nothing but ordinary. Going down, I passed through a pair of lead-painted metal doors, typical for the bomb shelters of the sixties. In my institute, a similar bunker was highly protected and was used as the headquarters of the regional Civil Defense. I walked through a narrow corridor, which after several turns fooled my sense of direction. I met a few people — nobody seemed to know where the Radiation Reconnaissance Center was. Wire-caged lights glowed dimly under a low ceiling; I saw shadows lurking along the dark-grey walls; I heard indistinguishable, muffled voices. The whole atmosphere brought the chills of a horror movie. Going through another set of two reinforced doors, I entered a large room, the size of which was difficult to assess due to the semi-darkness.

The air inside was stagnant and humid. The breathing was affected by the sharp, musky smell of sweat, mixed with something vaguely familiar from my chemical past — perhaps formaldehyde? Somewhere nearby I heard a low-toned, droning sound of a running motor, probably a diesel generator or a pump. A group of chaotically scattered wooden bunk beds emerged from the dark; I saw people sleeping on them. Guys back in the battalion told me that the most demanded personnel of the US-605 were stationed here: bulldozer, crane, excavator operators, skilled drivers, welders... all those who were always in great demand; those who already, according to the figurative expression of Igor, *glowed in the dark* due to a constant over-exposure to high fields, so they did not need the light anymore ... Probably, somewhere there was also that welder, whom I saw during my first jump.

Some lower bunks had sheets hanging from the top tier, for privacy; many were interconnected with a web of clothes lines — drip-drying underwear, towels added to the overall high humidity of the place. The electric shaver buzzed softly. I saw a man with an unbelievably pasty, ghoul-like face, sitting on the lower bunk and monotonously moving his torso sideways. He noticed me and quickly stopped.

"Cannot sleep ... Can't tell the difference between day and night, I live only from shift to shift. What date is today?" He asked.

"August the sixth," I shook my head and gave him a cigarette.

He immediately lighted it, not hiding.

"Doesn't matter ... Everyone here is smoking. After a few days in this crypt you lose the sense of smell, the air here is rotten, it's the bleach or whatever they use for disinfection," he said, rapidly blinking with teary eyes. I was sure he was not crying.

"How much have you got?" I asked the most common question among liquidators; everyone knew what it meant.

"About twenty roentgens ... " He chuckled sarcastically. "That's according to *their* records."

"So, five more, and that's it?" I was curious how it worked with the guys from *US-605*.

"Listen, man, why don't you beat it!" The phantom was suddenly angry. Cigarette, flipped in the air, flew to the corner and hissed, dousing. When I turned my head back to the ghoul, he had already slid into the darkness of his bunk bed, only his sharp knees and white clothed boots were in the spotlight in front of me. When I continued walking down the bunker, his last words were still reverberating in my ears.

"For us, *your limits* are inexistent. They keep us here almost chained, like slaves ... "

I did not know whether he was lying or not. The battalion pundits were saying that *these* were rarely replaced.

After a couple of turns in the maze of bunks and lockers I reached another conjoined hall. The light was brighter. On the next door there was the sign "Radiation Reconnaissance and Dose Control Group", and below a self-made sticker, saying: *"Viktor P., your flat in Pripyat was robbed! Check with Lagin, he knows more."* I sighed. Looting had progressively become a problem of the abandoned town.

Inside there were a couple of big desks with maps and folders. Sketches and maps on the walls were full of pencil marks, many times erased and written over. Two civilians slouched over one desk, head-to-head. A strong beam of a table lamp spotted a circle on the map, which they marked, looking into a folder and talking to each other quietly and quickly. At another desk I saw a silver-haired Lt. Colonel in a camouflage overall with the neck wrapped by a wide bandage. He spoke on the phone, holding the handset between the shoulder and the ear and simultaneously writing something in the notebook in front of him. Hesitant, I waited at the door.

"Varfolomeyev, you listen to me," the Colonel's voice was quiet, but stern. "When would you learn to quit bullsh*tting me? Your people are either liars or idiots, or both. Stop, I said, stop!" He stood upright. His voice didn't get louder, but I was sure that the guy on the other end of the wire did shut up immediately. "How much was there yesterday? One hundred and fifty? And the day before yesterday? Again, one hundred and fifty? So where in the love of mother Theresa did you get three hundred today, damn it? What do you mean, *the trucks brought dirt* *on tires*? What trucks? Who needs to come to your lamentable corner anyway?"

He finally noticed me and nodded to the chair at the desk. I sat on the edge, trying to be invisible. It was really interesting.

"Alright, I will get you two teams tomorrow, based on the levels of one hundred and fifty, and not three hundred, because if tomorrow the *fon* changes again, the people will just suck in extra radiation for nothing ... " He said, then took a pencil and marked something on the map. "Now, you have one more readout point with a stationary dosimeter at the junction with Nazarenko's district, right? Then you tell your dosimetrist not to be a bozo, he knows what I mean — there's no data from that device for the second day in a row, understand? Since I am mandated to collect the data points from all twenty-eight instruments, you give me this information by any means possible, got it? Dismissed!"

I noticed that the map corners were weighed down by ordinary jars with some kind of black powder inside. I couldn't believe my eyes; could it be ... a polymer absorbent?!

For several years, my department worked with one of the military-bound enterprises in Ukraine, co-developing novel polymer-based adsorbents for their uranium refineries, so I knew exactly how such a substance could be of use here.

"See, that's what the doctors recommend, and highly," Colonel said, pointing to the jars. "They are sure that it will remove radionuclides. One tablespoon a day, and this thing will pull out both strontium and technetium. Have a spoon, Lieutenant! My guys cannot swallow it without vodka, otherwise it feels creepy in the throat, like it gets stuck or something."

"Alcohol does not help, it will only make it worse," I responded.

"Where do you come from, smarty pants?" Colonel squinted. "How do you know this?"

"I know how this thing works, comrade Lieutenant-Colonel, well, at least, theoretically," I explained. "I'm a chemist. In general, one should take the adsorbent with water, and take on an empty stomach, so that there is maximum access to the walls of the stomach and intestines."

The officer looked at me with a slight suspicion.

"So, you don't drink vodka?" He chortled.

"Everybody drinks in the Army, comrade Lieutenant-Colonel. Since I am in the Army … " I did not finish the answer, making it obvious. "But I try to drink not too often … I drink according to circumstances."

"That's right," he said. "There are way too many here, who fill the tank to the brim, with all sorts of excuses and reasons … Good lad! What are you doing in our hell hole?"

I gave him an abbreviated report with a sketch depicting our *fon* measurements. He placed it under the bright table lamp, then looked at the big map.

"Well, you see, as I suspected, damn it!" He said, and his face darkened. "A few weeks of continuous erosion greatly diminish the surface readouts, but once you dig deeper, where the soil is still untouched from *the first days*, that's where the problem is buried! This definitely will be a big setback for weeks to come."

He raised his head.

"How long you have been here, Lieutenant?" He sympathetically nodded, learning that it was only my second time at the Station. "On Pikalov object, under the "Three", where did you hide, in the KSM building? And the sappers sat with you or in the Eleventh Canteen? That's OK if you don't know … But tell me this, do you know the rule *"two-four"*?" He said, and his eyes were sparkling with wit.

Something from my long forgotten military education came to mind, but not exactly. He explained to me that the rule applied to the calculation of radiation strength: with every doubled distance from the radiation source its level falls four times; for example, if I measure a 100 R/h on the distance of 100 m from the source, when I move further out by another 100 m, the level would be only 25 R/h.

"Remember this, chemist, it may come in handy," he said at the end. "Alright, hurry back to your quarters; maybe, we'll meet again!"

… The ghoul in the bunks vanished. Strange place, my peers were not lying about it.

I left the bunker with an odd feeling of getting used to the Station and its realm, bit by bit. The more I learned about it, the less I feared it. It was a reassuring finding.

Coming out to the fresh air, I received a shocker of bright light again; the sun was in its zenith, playfully bouncing rays off the bald head of Lenin's bust in the middle of the square. Unfortunately, sunglasses were not sold in the Brigade's mobile shop — responding to my request about that, the salesman sarcastically snickered: "Hey lady, would you like a bathing suit with your sunglasses, gift-wrapped?!"

Waiting for my team and *bort*, I watched with awe how a seemingly unresolvable jam in the square (several *borts*, two *BRDMs* and a mammoth-like *KrAZ*-mixer) was getting sort out. A soldier, standing behind the *KrAZ*, gave hand pointers to its driver for maneuver. The mixer had eleven mirrors installed along the perimeter of the cabin, where the driver watched the soldier behind. Still, the view from the driver's cabin was abysmal, regardless of the bunch of mirrors, since the holes of the leaded windows were tiny. It was not uncommon at the Station where these monstrous trucks ran over people, while backing up.

The jam eventually settled. I saw our *bort* approaching the granite stairs outside the hall where I waited, and a couple of minutes later my team came one by one from *ABK-2*. Everyone was accounted for.

Let's go home. The second jump was over.

We went through Kopachi's Radiation Checkpoint with ease. Unlike the infamous MM 00-02, that *bort* had not yet accumulated a lot of secondary radiation, either in the tire areas, or on the metal. Moreover, since Kopachi was close to the Station, its threshold was higher, 50 mR/h versus more stringent clearance of *Moldovan* checkpoint at the exit from the 30 km Exclusion Zone (5 mR/h). But the main thing was that the guys at Kopachi understood and appreciated our condition: exhausted, hungry, melting in the immense heat of the summer day. After a quick measurement, Kopachi's *dozer* waved us away.

We passed *PuSO-1* at Lelev with flying colors as well. All went much nicer and faster than the first jump. We were not going via Chernobyl, bypassing it to the right, through Zalesye. The road was worse, but the traffic was much lighter. It was amazing how quickly the day went by; it's already half past two. I felt very hungry, my stomach warbled like a bear. The canteen would probably be closed before we return. The alternative was a field ration, mmmm, yummy. I did not care, I would-take it.

Excitement gradually faded away. The driver understood my fatigue and kept quiet. I dozed off, listening to the monotonous squeaking of the seat springs.

In the evening I went to watch a movie in the Brigade "cinema theater". "Scaramouche", an old favorite of mine (the story was about medieval French noble/clown/avenger) promised to be a nice wrap-up of the tough day, but the copy was so brutally scratched and ripped that I felt frustrated and left early, heading to the Battalion Command Tent, which lately had turned into some sort of the "Officer's Club Past Ten": all my peers began to congregate in there after 22:00. It was probably the best time — to me at least — to feel the camaraderie, the elbow of the fellow next in line. Battalion officers flocked there in the evening; no drinking, no heated talks, just relaxed, casual atmosphere, chatting, laughing ... all that helped to get rid of the stress accumulated over the hard day shift.

There, I met Petrov and Igor, both chilling and smoking on the improvised bench next to the entrance. Igor informed me, keeping his typical boring facial expression and dry tone, that he would be replaced the next day by some Sen. Lieutenant Tkachuk.

The news saddened me. For those first several days, after the departure of Holodov, Igor had become my ultimate guru, a senior priest of the Station; he knew a lot and never skipped even a small chance to teach me more about the Station.

His miners had drunk their hearts out, giving a proper send-off to their commander. I said goodbye to Igor before midnight and fell asleep almost immediately, not paying attention to the hubbub in the tent.

The next day I had to take the first shift jump again, this time going to *"The Roof"* — rooftop of the "Three". Everybody was making googly eyes when I was mentioning this. Svyat said that "The Roof" was *"it"* at the Station — the most intense and fearsome radiation fields. I hid a smile in my mustache, which for some unknown reason had turned brick-reddish within the last couple of days (maybe, a humidity of *lepestok*, combined with its chemical disinfection components, had a certain bleaching effect). *We shall wait and see.*

... Waking up in the morning, I found a pair of sunglasses on the night stand, a goodbye gift from Igor. Shaped slightly orthodox, they nonetheless had wide and dark lenses, covering the eyes well.

I used them during each and every next jump, not paying attention to the woofing of senior officers, who were eager to see my eyes at the time of the berating. My excuse was bulletproof: Brigade doctors recommended me to wear glasses to protect the eyes while in close proximity to the source of radiation.

I shrugged off the commanders' nitpicking. I inherited these sunglasses from the high priest of the ChNPP.

PART 3

...Is Forearmed

August 7, 1986
Chernobyl NPP, third hodka

---

*"...Explosions have led to a complete
destruction of the reactor, its core, and
cooling systems, as well as the reactor hall.
Reinforced concrete and metal assemblies,
graphite blocks and fuel pieces were
propelled on the roof of the engine room (of
the unit 3 – SB) and were scattered on the
sizeable territory around the unit 4. A
strong flux of combustion products and a
powerful stream of gaseous radioactivity
emanated from a reactor crater, reaching
the height of several hundred meters..."
"...they were throwing the pieces of metal,
graphite, fuel assemblies from the roof of
the engine room by bare hands..."*
(From the site "Nefarious taint on the
history of mankind")

The wall of human backs in *propitka* in front of me was monumental,
like the sculptures of Stalin's times. All of us were synchronously step-
ping a couple of steps up the stairs and came to a standstill again. Heavy
breathing disturbed the otherwise quiet line of liquidators. Someone
stridently sniffed the air and broke the silence, chuckling: "Maaaaan,
you should stop eating cabbage for breakfast! I am afraid to light a
match here!" The line responded with laughter. However, puns and
jokes quickly faded away. The stale, motionless air felt rigid. Smoking
was not allowed, but people didn't care, judging by the continuously
strong smell of smoldering tobacco.

The mark "14" on the wall.

It looked like this line would take about three more hours to
ascend ... But I was wrong.

Every *hodka* to the station was an event, a big deal for the liquidator.
The jump to "The Roof", where the liquidators were removing highly
radioactive debris from top levels of the "Three", was the apotheosis of
everything that I saw at the Station.

Both reactor units, the third and fourth, were structurally positioned in the same building, under a single roof. One could climb to the roof of the "Three", located at a height of over 70 meters, wander a bit through a maze of transitions from level to level, and then, using the red-white striped deaerator stack as a landmark, approach the exploded "Four" from the top and even look down into the crater.

If someone really wanted to do that.

We often argued with battalion pundits: would such a brave man have enough time to descend on his own? Maybe.

The first firefighters did go down themselves.

Those who did were already buried at a depth of six meters in lead coffins.

"The Roof" was of difficult terrain, highly uneven, with multiple horizontal sections, cut-offs, platforms and levels, interconnected with flimsy fire escape ladders. There were several exits that were considered for bringing troops as close as possible to the place of work, but almost all of them were rejected due to the atrocious radiation outside the attic. The radiation landscape and levels were constantly changing: atop of the insidious nature of extremely radioactive projectiles, initially thrown out by the explosion and then cocooned in the molten bitumen on the roof (I return to this problem a bit later), the intermittent microexplosions in the crater varied the levels in great margins. Such unpredictability had made "The Roof" a very stressful and unpopular job among the liquidators. The highest radiation fields (in some places, up to several thousand R/h) were recorded over there. But we had our orders.

In my first jump to "The Roof", I witnessed the work of a rover robot, which had been purchased for some crazy money in West Germany for unmanned remote cleaning of the roof. There were three or four civilians, gathered near the exit on the roof inside the attic, who were trying to move the robot about the roof section filled with debris, using remote control. The picture that was received from the robot camera was shown on the monitor; it was fuzzy and distorted. For a while, the rover was moving rapidly and seemed to follow the joystick of the remote control, but soon, after a few erratic moves, it stopped, and the picture eventually went out. The high reputation of German telemetry was proven powerless in treacherous levels of radiation.

One of the factors that greatly complicated the qualitative and efficient cleaning of "the Roof" was that a large number of fuel element pieces, graphite and chunks of fuel were scattered on the roof of "Three" in places, covered (in a violation of building codes for nuclear plants) with a thick layer of the ordinary bitumen. When the fire started, bitumen melted from the heat, and extremely radioactive solid pieces sank into the liquid tar as in quicksand. The bitumen then solidified. No traces of the "drowned" pieces were left on the surface. They had to be detected like truffles, almost by the scent. Only instead of trained pigs, experienced *dozers* came to the rescue. They marked the identified places of drowned solid radioactive pieces by sticks and small flags.

When I saw it for the first time, and heard the briefing given by the civilian customer to the team, the movie "Stalker" came to mind (*"If you doubt whether it is safe to step on that patch or not, you throw the lug nut, then wait and watch: if it disappears, do not go there!"*) The feeling of incongruity, the movie-like oddity of what was happening on "The Roof" did not go away even after the third or the fifth jump. Just imagine the equipment we wore while working there: in addition to *propitka* and helmet, we wore the custom-made lead "armor" that was covering the torso: two thin lead plates, tied together with a wire, shielded the chest and groin in the front, and back and buttocks behind. The face was protected with dustproof glasses and a heavy respirator RPG-67 with two side cartridges. In my first *hodka* to "the Roof", we also wore shoe covers from OZK, which, just as the respirator, were shared — there were not enough of those for everyone. Finally, thick welder mittens were put on the hands. Just to walk around in this "protective outfit" was already a daunting task, forget running or working in it.

When I finally ascended to the top, I feverishly gasped for air, looking like a fish yanked out of water.

One wall of the attic exit cube had a three-dimensional map of "The Roof", drawn extremely skillfully with chalk and brick chunks. The customer, a civilian, stocky, dark-haired man in his fifties, dressed in white overalls, either from Kurchatov institute or from *Minatomenergo*, quietly but clearly explained the task to the group of soldiers.

I went first from my team. That was my way of facing the edge; I hated time lags. Plus, I truly wanted to quickly end that anxious anticipation of the unknown. Just like sitting in the dentist office. You wait, and you wait, you're nervous; too much time seems to be stretched like a slinky spring. You are not called in and you just badly want to get rid of the damn tooth *right now*.

...In order to understand better to what extent the process of "ascending to The Roof" affects the human psyche, think of the following picture. Imagine a stairway coming up from the entrance to the roof of a twenty-five-storey building. This stairway, zigzagging back and forth, eventually ends in a small attic, a caponier, located somewhere at a height of 70 meters, with a door leading to "The roof". During the whole route up, there were no windows, no doors, just a dull, monotonous ascending: fourteen steps... platform... fourteen steps... platform...

Ad nauseum.

On the platforms, there were big numbers, indicating the so-called marks, or levels (storeys), painted in red. A few dusk emergency lights on the way up did not dissipate the darkness, and certainly did not lighten the mood of the line. Adrenaline heavily spiked your brain just by acknowledging the fact that you were going to "The Roof", which loaded it like a gun with anticipation, tension and nervousness.

On about fifty flights of that stairway and on all platforms in between them, stood the line of liquidators. The line did not proceed fast, on the contrary, it moved slower than a snail's pace. Some soldiers tried to sit down on the stairs, but it was not easy because the stairway was packed with liquidators, front to back and shoulder to shoulder with the fellows next to you. The line was packed solid. All stood quietly, patiently. The length of your wait depended on how quickly the guys on the very top finished their shift. Repetitive, submissive upward movement, one or two steps each time, then another stop. The lucky ones, who were done, were literally roller-balling their way down, pushing aside the green ribbon of bodies who were nervously sweating in the queue.

I had plenty of time for the math exercises during my first jump to "The Roof". It appeared that about six hundred people were languishing in the queue, even by the most conservative estimates.

The absurdity of this otherwise noble and colossal task was boosted by the fact that all of the liquidators who were scheduled for "The Roof", had to arrive for *razvod* at the same time, which always started at 08:00. Then all teams for "The Roof" were coming to *ABK-2* for change of clothes, and were reaching the stairway, which was located inside the joint building of the "Three" and "Four", pretty much at the same time. This created a dilemma: if you were more fortunate, or more prudent, to get to the levels above mark 20 before the shift commenced, then you hit the jackpot: in about thirty minutes (one hour tops) of edgy waiting you would reach the attic, and after shockingly quick work shift you would be done.

However, if not …

In the first jump on "The Roof", we hit the tail of the queue. *We had to wait for over five hours.*

So, in a way, this stint was more about a race in the beginning of the shift.

… The heavy pulse rattled loudly in the temples and ears, meddling with the briefing of the customer. His voice was barely breaching the veil of adrenaline.

There it was. The deadliest. The most intense.

I could hardly digest what he was talking about, but I *must* listen, because he was talking to our team. Our lives were on the line, like at war … and I couldn't concentrate!

Alright, the maximum allowed time to be on level … What did he say — "L"? "N"? Or what?

Hold on, he said how long? One minute? Did I get it right?!

Holy mother of God.

Five hours. And one minute. *Sixty seconds.*

Let me re-phrase it again: waiting your turn for five hours and working out there, on "The Roof", for sixty seconds.

Of all grotesque and purely irrational things (albeit often deadly in a full sense!) that we were doing at ChNPP that hot summer, this is the one that will never fade in my memory (one of my next jumps on "The

Roof" had lasted only 35 seconds!). I realized this much later, but for now ...

I finally got a grip and began to absorb the words of the "Roof Apostle": get out of the attic, go straight to the wall over there, climb one level higher, turn around the corner to my left (follow the marks!) ... Good. Then ... Damn, what then?! Do I have to go up one more level? Yes, thanks (he pointed out on the map, where we actually had to work). I am good with maps, and it looks like I've got the path right. Alright, I am there; I have to grab a makeshift "ice axe" (an axe welded to a long steel rod), chop a few slices of bitumen, then pick up these slices with a shovel, run to the roof edge and toss them down. Repeat if I have time.

Looks trivial.

I realized how difficult it was when I started jogging (it was impossible to run in that "protective suit") to the fire escape ladder that supposedly had to takeme to level "L". He pointed to it on the map, I remembered that, but the mark on the wall next to the ladder — an arrow pointing up — had a letter "N" next to it! *Or is it this darn ladder around the corner?* I couldn't breathe. *Focus, focus!*

Short spurt around the corner — and the ladder was not there. *Where is it?*

My heart was ready to jump out of the ribcage, knees were buckling under the heavy weight of lead plates. For a moment, I recalled my basketball practice times, when we ran in the drills back and forth across the court, holding 20 kg weights in hands.

Stop.

I saw another mark, pointing around the other corner to my right. Intuitively, I followed it — and there was that lousy ladder! I climbed frenziedly, slipping once or twice, drenched in sweat and suffocating in the respirator that seemed clogged, because I couldn't get a full breath that I desperately needed. I began counting once I stepped out in the open, in the manner of "one Mississippi, two Mississippi ... " (*one milliroentgen, two milliroentgen, three ...*), but very soon dropped that idea, because counting was plainly shutting down my focus.

My sports-trained legs, thank God, seemed to function separately from my panicked brain. In two–three seconds after my climb, I reached another ladder, shorter but flimsier than the first one. I grabbed the ladder railing, but at that very moment the guy, whom I had to replace, literally jumped down, barely touching the ladder and almost landing on my head with his heavy behind.

I couldn't see his face — my goggles were misty from the hot sweat inside, I couldn't see small details. He looked just like an alien. Together with exhaust air, he pushed off through the membrane: "The shovel... There... Roof railing...", waved his hand in the direction of the next level and ran to the ladder down.

I climbed up like there was no tomorrow. Legs were ready to give up. The suit was soaked in sweat; I felt as if the shoe covers were filled with it and were squishing with each step. When I reached the spot, I didn't waste a second to look around. There was the axe. *Grab it, chop it. Chop it. Chop it*?! Stubborn bitumen had turned into resinous mass under the sun, and resisted my hits. I continued hacking madly, and with each hit I felt like my brain boiled with rage more and more. *Come on, you stupid thing!* A slab of bitumen finally separated; the second one went faster, so did the third one. *How long am I here*?! I hated the cliché "It seemed like an eternity", but over there I *physically* felt every second.

Someone pulled my sleeve — my replacement came up already. *Was it really only a minute*?!

I dropped the axe and picked up the shovel — all three bitumen slabs in a moment flew down over the railing, and I scrammed back.

I don't remember the way down. I recall that I was close to breaking my neck, almost falling off the last ladder. In the caponier, pouring sweat out of the respirator, I yanked the lead armor off me and leaned against the wall, completely exhausted, feeling the chills racing up and down my back, like waves. My head was spinning. "It's a blackout", I thought, and with this thought I departed into something warm and soft, like a cocoon.

"Lieutenant, shoe covers! Take off the shoe covers, you hear?" Someone was shaking me.

I pointlessly stared at the belt buckle of the soldier, who was pushing my shoulder. The sickle and the hummer over the star on the buckle. How many decades this symbol hadn't changed?

"C'mon, man! Damn … I am next, I need shoe covers!" The soldier's voice was getting impatient.

I nodded in understanding. Fingers did not follow the brain's commands, slipping from the rubber fasteners. Sounds were muted, as if my ears were filled with cotton wool. I finally took off the covers.

The civilian gave me a cigarette.

"You are done for today. Chill out a little there," he pointed to the stairway. "And then dash to *ABK*."

"How … " I coughed several times, heavily, then forced myself to stop. "How long I was … "

"Minute and a half," he said, and quickly added to calm me down. "I saw you running around like a headless chicken for a while, before going on to the *high field*. Don't sweat it … I mean, really. You'll be fine!" He chuckled softly.

I barely stretched my lips into something like a catatonic smile and slowly walked to the stairs. Holy cow. That was something.

I passed by my troops and set a rendezvous time in the lower hall of *ABK-2* after deactivation shower. They didn't need me there; and even if they did, it probably would take some effort for me to act in full capacity at the time.

Going down the stairs was tough. The legs felt rebellious and with every step wanted to go about their own business. I noticed that the lower part of the staircase was empty. The last guy in the queue looked at me with such envy that I stopped for a second and gave him a cigarette, which I got from the civilian upstairs.

After that, only the clatter of my heels was breaking the sacred silence of the stairway. It was spooky. In the semi-darkness, in the damp, foul air I envisioned an endless chain of shadows in the green uniforms, wearily rising to the carnage. The owners of these shadows were gone, leaving behind a stench of tobacco smoke, disinfection from *propitka* and sweat, heavily mixed with fear.

Who says that the way to hell is heading down?!

August 13, 1986
Chernobyl NPP, ninth hodka

> *"... The long-term stress, the chronic impact of continuous external and internal exposure to radiation, the lack of scientific knowledge related to how the multiple factors of radiation exposure (biological, chemical, psychogenic aspects, etc.) influence the human health, the absence of information and reliable means of protection — all this predetermined the deterioration of health of the liquidators."*
> (From the site
> http://stopatom.slavutich.kiev.ua)

Steady rain lazily slapped my back with long whips of surprisingly cold water; my reefer jacket was already drenched all the way through, to the *propitka* underneath. I physically felt water trickling down my skin. All hail the marvel of resilient Army technology, "impregnated uniform", darn it.

The short hat's peak couldn't protect the cigarette from drips; it hissed with anger and broke apart as soon as I touched it. I cursed. Hands were freezing in the wind, I kept them in my jacket pocket. The reefer was soaking wet and put a ton of weight on the shoulders, but it still was better than nothing, at least some protection from the dank rain.

We stood a few meters away from the truck crane, lying almost on its side, overturned by the weight of a large, two-by-eight meter, concrete slab. The crane missed the target, a new, three-slab wide, paved road that we were building from these flat behemoths: the crane dropped it on a pile of pebbles on the roadside when, due to too heavy load (that slab, for some unknown reason, was visibly thicker and more massive than the rest), its right side began to slowly rise in the air.

The auto-grader, which was preparing the pebble bed for slab installation, coughed a small cloud of blue smoke from exhaust and stopped. Pebbles were clearly unfit for the size of monstrous puddles, across which we needed to lay slabs, but since we had been ordered to lay out a new road, new road it was. I looked with regret at how a huge pebble hill was sucked up by slushy mud, like granulated sugar disappears in hot water.

For a week, the liquidator teams had been on pins and needles, encircling the Station with a guard belt, *"betonka"* — the road made from concrete slabs. The road was surrounded on both sides by a barbed wire fence. It had to be built alongside the existing dirt road, that was spontaneously made by the drivers of countless machines on wheels, which were populating the Station by the hundreds in those days. A new Brigade Commander had arrived. One of his first actions, rumor had it, was a report to the *SGC* that the security level at ChNPP was intolerable, beyond comprehension. I didn't really know if the guard belt would really help, but the point of the matter was that all teams from our Brigade were building it non-stop, three shifts a day.

Last week, the US-605 had finished digging a super deep narrow trench, which separated the Station from Pripyat River, and filled it with a bentonite clay to build what was coined a "wall in the ground", therefore eliminating a threat of radionuclide penetration into the Pripyat aquifer. According to the word of our battalion *analysts, Gos-gidromet* immediately stopped dispersing clouds in the vicinity of the Station. Almost two months of unbearably dry hot weather came to an end. About the same day, in accordance with all laws of cruelty and injustice, a non-stop rain started to pour, and it literally had only a few hours of break ever since.

The liquidators cursed the work in the open air of the Industrial Zone. By mid-August, the *fon* on the outskirts of the Industrial Zone, in the areas away from the "Four", fell to 50–100 mR/h, sometimes less, which meant that even the full shift (8 hours) did not cover a daily quota of 2 roentgen.

But the dangerous *vtorichka,*[1] the huge unknown factor of absorbing tiny treacherous radioisotopes from dust, mud, rain, even from polluted air, led to the accumulation of radionuclides in the lungs,

---

[1] Secondary radiation.

stomach, kidneys, liver, hair, skin … it was a ticking time bomb waiting to explode in the distant future in the form of cancer, abnormalities, other health issues of all liquidators.

Who can tell now for sure, how many liquidators already lost their lives and how many will succumb due to the *vtorichka* in the years to come?

Places on the Station, where liquidators were called to work with "direct" (penetrating) radiation — a combination of neutron radiation and gamma-, beta-, and alpha-rays, were more desirable to work at for many. The same guy, a civilian customer during my first jump to "The Roof", led our work during my second jump there as well.

"People are afraid of radiation far more than of fire," he said after my second jump. "Maybe this is right. However, here, when you work on "The Roof", the penetrating radiation is *clean*. It's like this: you caught a *"butterfly"*, several dozens of roentgens… Too bad, but you've got medical help, hopefully, your blood goes back to normal, and you are ready to go again. I received a couple of times like 50–70 roentgens each, maybe more; still alive and kicking, as you can see. Far worse thing is to work in the polluted, dusty and dirty contaminated area when you swallow or inhale the radionuclides that accumulate in your body, like your pension plan in a bank. You went to "The Roof", you worked for a couple of minutes, and off you go, clean and happy … But there," he nodded down towards the Industrial Zone, "you swallow and breathe the isotopes while you puke your guts working for hours on the surface decontamination, and then these isotopes calmly sit, you know, like truffles, in all your internal organs from stomach to bones, and steadily deteriorate your health, decomposing with a half-life of a few hundreds of years!"

That guy was a good man. I saw him once or twice after second jump to "The Roof", but unfortunately, I don't remember his name. I still regret that I did not take my Station notebook back with me.

"The Roof" jumps had added a new detail to the set of odd reactions of my body to radiation fields of varying levels. Immediately after each of my six jumps there, a severe headache was mangling my head, my nose was getting clogged so viciously that I had to periodically remove *lepestok* to catch a gulp of air with my mouth on the way back. Three or four hours after every jump, the symptom disappeared without a trace: the headache was gone, the nose was breathing normally ... and that was it.

Dazed by the observation, Svyat called it an allergy to the super-high radiation fields.

I tried to light a new cigarette. Before I was able to get the first puff, it broke in my shaking wet hands. Hard to believe that it could be this chilly in August! Scratching in the throat was getting worse. *Hello, flu. All I need right now.*

The "expert panel" near the truck crane consisted of four officers: two inspecting Colonels from the Brigade HQ, plus a Captain from the *OGMO*, plus a very active Senior Lieutenant who had arrived recently in a personal *BRDM* (must be a KGB guy!). The two Brigade Colonels were very edgy: the new honcho, Brigade Commander, had made such daily checks of the teams that worked at the Station, as mandatory for his "top dogs" — they were not getting their dose level records filled for nothing anymore; now, to get their roentgens, they must at least show their sorry faces at the Station — not on "The Roof", naturally.

Next to them, timidly hiding his head in the wide collar of his reefer jacket, stood the repentant crane operator. The "panel" gave him a hearty tongue-lashing, for good cause: when he dropped the slab, it splashed some serious mud on their brand-new overcoats.

A set of generous wishes for the crane operator's mama were followed by a meaningless squabble between the "panel" members on what to do with the slab and the crane. I was farsightedly silent, standing between the panel and my team, which was not happy with how the events were developing — I sensed their frustration. The most logical idea out of the visitors' "brainstorming" was the suggestion to pile up everyone on the dangling edge of the crane and "yank" it. Maybe it will go down. "Maybe it won't ... " muttered one of my troops. I gave him the warning look. Ten soldiers, soaking wet, exhausted, hungry and weary, clearly were not enthused to follow this iffy idea. Anything could happen, and this wasn't the brightest plan from the safety perspective either.

I braced myself up. The storm was nearing.

Meanwhile, the damn slab continued sliding down the pile of pebbles, slowly pulling the slings that tethered it to the crane arm; the slings were making a depressing groaning sound.

The rain intensified.

I swear that right there I truly understood the meaning of the expression "to cut the tension with the knife".

The circle road around the Station (if one could call this beaten path, in ruts and gullies, a road) was sparsely populated during these rainy days. Every stubborn driver — and those were plenty — was still trying to cut straight through the Station, often without realizing the possible dire cost of such a bold drive. During our jump, seemingly endless, only three or four trucks and specialty vehicles passed by us, not counting those bringing slabs and pebbles for the padding.

As in a fairy tale, I suddenly heard an engine noise approaching, and in a minute or two on the circle road showed up … another truck crane.

Why did the poor guy choose the circle road and where was he going? That didn't matter to the "panel". They all jumped in front of the crane, pulled out the totally dumbfounded driver and started barking in four voices, that regardless of his previous assignment he was *commandeered* to help out here. The other crane was definitely more powerful, with a telescopic arm, but even to my modest knowledge of physics laws it didn't look strong enough to lift the darn slab, which now almost overturned the smaller crane.

A possible solution hit me rapidly. We didn't have to use the new crane alone; it could help by distributing the load between the two cranes!

The panel, scratching their heads, reluctantly agreed to give it a try.

Sizable time was spent on useless and partially hysterical instructions of each of the "panel experts" to both crane operators, but this was the price tag of having the "panel" on-site. Finally, the stronger crane hooked and picked up the top end of the other crane's arm, and together, revving and trembling, they dragged the stupid slab in place on the pebble bed.

In all this turmoil, our time had expired, and — thank God — the replacement team showed up as expected. Deliriously happy, my troops sloshed to *ABK-2*. I told them not to wait for me, because I was graced by a personal invitation to shower at *ABK-1*, which was used by the Station personnel, the Kurchatov Institute and the *OGMO* guys, and where normally we, simple folk reservist liquidators, were not allowed to change. My new formal commander at AAS unit, "some Sen. Lieutenant Tkachuk", had started daily trips with our mighty miner team to the Station laundry, which was located under *ABK-1*, and asked me to come over after my shift, so I had a legit reason to go there.

Tkachuk, who replaced Igor, was a Western-Ukrainian guy: short, wide-shouldered, with flickers of incessant out-of-town wit in dark-brown eyes. His roundish, black-haired head was somewhat disproportionately big, which gave him an excuse to wear the white cylindrical cap of nuclear operator even at the Brigade, based on his sworn statement that there was no appropriate size of military hat at the *perevalka*. He was almost my age, and before *the war* worked as a chemical engineer at one of the chemical industry giants in the Central Ukraine, synthesizing indigo-based dyes; this built a quick profession-based bridge between us. His first name, Evgeniy, was shortened in his home town to "Genyk", and that's how he introduced himself to me.

Genyk had quickly gotten along with the miners. In our tent, he occupied a bunk next to me, and before going to bed, often entertained me with colorful stories from his civilian life, spattering sentences with a cute interjection "*ooot*" (the Ukrainian analogue of the Russian "*vot*", popular word-parasite), which gave his speech a distinctive charm. He also used a heavy mix of both languages, which did not bother me as I was fluent in both.

When I met him at the *ABK-1* entrance, Genyk was dressed just like Igor was, in a blue Station technician uniform, with a traditional white cap.

"It's so hot in the laundry. Though we change into the light clothes, but still, *ooot*, it's too sweaty," he said, walking me up the stairs to the third floor. The shift change ended, and the building was empty. After nearly eight hours in the rain, my chilled body gradually warmed up. Fatigue suddenly hit me like a wave.

There was no one in the changing room. I dropped my soaked reefer jacket, uniform top, shirt and trousers. I was so tired that it took me some time to take off my boots and socks. My feet were buzzing with cold. Genyk supplied me with new bar of soap and — can you believe it! — flip-flops. I went to the shower.

Eternal bliss. Hot water, and I meant — hot. Skin-peeling strong jets. Aromatic strawberry soap. I reminded myself of a sybarite.

After the shower, I felt like I could eat a cow. While I was changing into a clean uniform (here, at *ABK-1*, they carried only white or blue uniforms and white boots cut off at the ankle), Genyk smoked, puffing in the half-open window.

"Let's go, *oot*, we'll eat in the main dining room, I have an extra coupon," he said. His hospitality had no limits. The reservist liquidators were not allowed to eat in the Station canteen, but Igor told me once that the civilians ate like czars compared to us, Brigade rats; they had quality supplies and real chefs, which meant that their *borsch* was the real deal, not dishwater with sparse rotten cabbage leaves and a thick layer of melted fat on the surface.

The coupon was not required. The dining room was empty, and while seeing two good-looking "Station technicians" in blue outfits, the smiling girls behind the counter piled for us freshly cooked borsch, aromatic buckwheat porridge, and cutlets made from real meat. A pickle, a tomato and two glasses of apple juice were the worthy add-ons to the feast. I almost cried, when we sat down and began devouring all that. I felt that I had not eaten such a discerningly good meal for several years. The Brigade chow was recalled with a shudder.

The rain had stopped a while back. After a hot shower, a hearty meal, and another change of clothes, the sensation of a cold was gone. I still wouldn't mind getting a *clean* reefer jacket, but they were not distributed at *ABK-1*, and I did wear three shirts under the cotton jacket of the reactor operator. Waiting for my ride, I smoked a cigarette with Genyk, standing at the main entrance to the Administration building. The wind was ruffling the edges of hand-written notes on the billboard. Most popular ones said: "*I'm looking for ...*", "*If anyone knows ...*", and variations like "*Tolik Klebanov, we were moved to Kiev on Goncharov St., Bldg No. 9 Flat No. ...*", "*Valya G., if you are here, please find me in the pioneer camp "Skazochnyi"...*". Very often I saw the word "*war*" — that's how the locals called the accident and the aftermath events. "*I'm looking for Victor D., before the war he worked as a diesel mechanic of Unit 3…"*

The Wailing Wall of Chernobyl.

I shrugged my shoulders. That was really unsettling.

Unfortunately, there were no *borts* coming out of the 30 km Zone, and for a while. Genyk went to the laundry a long time ago. It seemed like an endless wait in the hall of the Administration building. The rain started pouring again. I became desperate, and, upon learning that the team, which was loading at that moment in the central square, was going to Chapaevskaya village (which was somewhat afar from Oranoe village where my Brigade was stationed), I asked for a ride.

I sat at the window in the cabin. The driver and the team commander, the young Lieutenant, looked respectfully at my Station uniform. Despite the three shirts that I had under the thin cotton jacket, I felt cold and lifted the collar. The driver noticed that and turned on the heater. I nodded in acknowledgement and closed my eyes. Falling asleep under the mixed droning sound of the motor and the rain, I thought that I began to get used to the Station; I felt more and more comfortable with the totally different mode of my life in the last… days? Weeks? Months?

Does it matter?

Willingly or not, I was fitting myself into the incredibly complex existence of the Chernobyl nuclear power plant. The Station did not scare me anymore; at some elusive moment it became a part of me, and in a beholden trade I was somehow, slowly, becoming a part of it, just like hundreds and thousands of others, dragged by the edgy hands of the "Four" reactor operator into something the planet had not yet experienced.

August 17, 1986
Chernobyl NPP, thirteenth hodka

> *"... The efficient methods of decontamination that were developed by chemists of the Academy of Sciences of Ukraine (A.A. Chuyko, A.V. Chuprin, L.I. Rudenko, and others) in the first months after the accident made it possible to reduce the concentration of radioactive aerosols in the crater of damaged Unit 4. Chemical compositions and methods of dust suppression and soil adhesion were also successfully put in practice (V.P. Kukhar, V.V. Blagoev, V.G. Sklyar, etc.)."*
>
> (From the site
> http://stopatom.slavutich.kiev.ua)

Gorin unenthusiastically scraped a piece of polymer film off the concrete. He joined me again, for the third time, and I somehow felt safer knowing that he and his blue-taped *depeshka* were here. The size of the detached film portion clearly did not impress him, and me as well. I looked with anguish at the vast backyard of the Administration building, fully paved with concrete slabs and covered on the top with a film of dust-binding adhesive polymer. Our task was to peel the film off the slabs and sprayed them clean.

The idea, which once seemed brilliant to many, eventually turned into a headache for the Station. While aging (and it happened relatively quickly, within the first couple of months), the film became brittle and was gradually ripped off the slabs by the tires of vehicles in traffic that was darting non-stop through the Industrial Zone. Soon the film was partially peeled off, forming rag-like small "tongues" that were however still gripping the concrete. Radioactive dust, spread by the wind, was trapped under these "tongues", creating a nuisance of relatively small but steady radioactive contamination around the central part of the Station.

As in a sarcastic old Soviet saying: "In this case, we aimed for the better, but it turned out as usual, which is terrible".

To our disappointment, the film did not want to part with concrete so easily. Fifteen troops from my team kept scraping the polymer with shovels, and it did not look promising at all. Strangely, in many places the adhesion of the film to the concrete was way too strong for a simple mechanical rubdown removal. I timed the team efforts: about 10–12 square meters in half an hour. I eyeballed the humongous backyard. *Jeez, we will be here until New Year!* The Colonel from *OGMO*, who was our customer today, cheerfully said at *razvod*: "You have to quickly remove the polymer layer, and then the firemen — you get one engine with you — wash off the dust, and that's it!" Goodness gracious, did you at least try it, before requesting only 15 troops for this monkey job?! I almost spit in frustration, but stopped in time, realizing that I had *lepestok* over my mouth. Not even a day, not two, much longer. Unless … I scowled and looked at the firefighters' Chief, a brave-looking guy (he had a huge black mustache with twisted edges) in nicely fit clean uniform.

"Well, I can try and blast off this cr*p," the Chief said, as if he was reading my thoughts. "However, there is not enough water in the tank, and the sprinkles will use it in a hurry."

"Let's give it a try," I replied. "We are not losing much, we get diddly with shovels here. How much time do you need to refill the tank?" I tried to figure out where the firefighter could get water, and nothing smarter than a cooling pond came to my mind.

Hesitant (I suspected that radioactivity of water in the cooling pond was rather high), I still shared this idea with the Chief; it turned out that they already filled their tank from the pond, and not once. It was slightly radioactive, but the *fon* there was still higher anyway. The problem was that it would take about half an hour to refill a full tank — the truck's sump pump was too slow.

So, I calculated, together with the way there and back, about one hour just to get the fire truck ready. Not really helping.

The Chief and I edgily watched how the soldiers, swearing, scratched the slabs. Radioactive dust, raised up by the shovels, hung in

the air, increasing our exasperation. I physically felt how the dust pene-
trated through the fibers of our respirators, which were getting ineffective because the longer we wore them, the wetter from the breath they
became. Then the radionuclides settled in the lungs. Maaaan...

Nonetheless, we started.

The engine frighteningly roared for about five minutes and stopped.
Only about three or four square meters of concrete were cleaned, but
I had to admit that the quality of removal was far better than with
shovels. Plus, that was water-assisted cleanup, which meant that there
were no radioactive particles elevated in the air.

Fine, we do what we can.

While the fire truck went to the cooling pond, I sent my men to
the nearest *otsidka*. I, however, should be on the lookout in case the
Brigade inspecting officer showed up. Since I didn't want to hang
around in the open, I tried to find anywhere to hide. Then I saw it; near
the corner of a small technological building, about 50 meters away, was
an abandoned "Ural" truck, stripped for parts almost beyond recognition. It was left here in the fever of the first days, and was gradually
mauled by car thieves. After losing tires, it was immobilized, and then
anything that could be removed with screwdriver, wrench and hammer,
was stolen. I opened the cabin door and turned on the *depeshka*, which
I borrowed from Gorin. To my surprise, it showed only about 40–50
mR/h inside, roughly four times less than out on the grounds. The wind-
shield and door windows were still intact, helping to keep radiation in
the cabin lower.

The seat cushions were stolen a while back, apparently. I tried to
find the best position to get comfortable on the metal floor. Nice. Quiet,
safe, relaxing. I closed my eyes, pulled down the *lepestok* and lighted up
a cigarette. The fatigue clung to my spent body instantly. Over the past
few days, I had gone dog-tired and fidgety.

Take that morning, for example. When we were about to leave
from the Brigade, a young Lieutenant, in mint condition uniform and
squeaky-clean boots, ran to our *bort* and waved his hand, demanding
to stop and open the door. I rolled down the window.

"Comrade Senior Lieutenant, how many Communists and Komsomol[2] members are there in your team?" He asked with stern voice, preparing a small notepad and a pen to write down the numbers.

My first reaction was to chuckle, but I knew right away that he was not joking. I looked at his pristine clean uniform, snow-white undercollar, notepad, his face full of pride and sense of importance ... And I snapped.

"Listen, pal, were you born yesterday? Has anybody explained to you how the teams are compiled here? Of those who are there, in the back of the truck, I barely know two or three guys by their face, forget the name, and surely, the last thing I know about them is their party affiliation!"

His eyes turned white with anger; he obviously was one of those Brigade HQ clowns who spent all their days in a huge clean tent with desks and fans inside, knowing about the Station only from the media, despite being a few kilometers away from it. He was fuming, and now – if he wanted – he had a reason and a capacity to squeal on me, but I did not care. Tiny devilish bubbles of cynicism were boiling in my blood, knocking in the brain, and I released them as quickly as I can.

"Here's what we can do, if you really want to get these numbers. Jump in, with the team out on the back, talk to them until we reach the station; better yet, you work with our team in the Industrial Zone — to get familiar with us, simple folks. They say it's only about two hundred milliroentgen today out there. The shift is long, so you will have plenty of time to get to know the guys closer, eh?"

I slammed the door in his face, and we left.

Was I expecting to be penalized? You bet; however, even if he pawned me, what would they do to me? I was already in the worst possible hellhole on mother Earth! For them, "The Roof" was the worst thing, for me — the best possible scenario of what was left to do at the Station.

I almost forgot about the morning incident, and it started aggravating me again, so I decided to focus on the job.

---

[2] (Russian: *Vsesoyuznyi KOMmunisticheskiy SOyuz MOLodezhi*) A youth organization across USSR, with close ties to the Communist Party.

Alright, the bottleneck is the slow feeding of the fire truck tank with water. What if we have ARS instead of fire truck? But ARS sprinklers are of low power, they are not designed for releasing a strong spray, individually they are far weaker than those on the fire engine. However, is it possible to add... Yeah, that's right! What if we would have a tandem of ARS and the fire engine? ARS would be used as a water feeder for the fire truck, which would strip off the film. By God, this should work — and to feed the water from ARS to fire truck tank would not be a problem, too! But where to get the ARS? If only I could find that Colonel-customer from the morning...

Excited, I jumped out of the "Ural" cabin. My legs responded with an ache in the joints, like I had the flu. I was reaching the limit of my physical capacity, and that was a snowballing effect, for sure.

I told the troops to stay put in the *otsidka* and went searching for the Colonel. I found him after a brief meandering through the corridors of the Administration building and enquiring in several offices. Surprisingly, the Colonel was better than I thought of him initially. Less than an hour later we already had a couple of *ARS* assigned from our Brigade's First Battalion of Special Operations, and we were back in business. All that my guys needed to do then was only to shovel the washed off wet strips of film into the heaps and then load those on a dump truck.

At the end of the shift I looked at the freshly washed backyard with well-deserved satisfaction. It's time to hit the showers, but right at that moment I saw a *BRDM* with a familiar number on the armor that was turning around the corner of the Administration building and heading our way. Good grief, Battalion Commander Golubev, no less!

A week ago, following the trend of the Brigade top dogs, he introduced the practice of "sudden raids" at the Station for him and *COS* Zelenov. These visits were needed to justify their dose accumulation — the sacred 25 R limit was written for them, too. Thus, they were rolling here and there, following our *BRDM* crews at daily reconnaissance trips, coming to Genyk's laundry, chasing me around the Station... Of course, not courageously jumping into the devil's teeth, but close enough, in their estimations.

BC Golubev, red-faced from sucking in his guts (the *BRDM* hatch was too narrow for his girth), squeezed himself out. Gossiping in the "Officers Club", we were wondering why both our seniors did not lose much weight, as all of us did. The conclusion was that both Golubev and Zelenov were huge fans of "sip-and-bite" exercise, and that they apparently had a stash of tasty food supplies — and vodka! — in their tents.

After a traditional short paternal talk (it sounded something like this: "How are you?" — "Thank you, Comrade Lieutenant-Colonel, I am fine!" — "Are there any problems, anything that I can help you with?" — "No, Sir!" — "Great, I trust that you report the shift summary on the operations meeting, then!" — "Yes, Sir!"), BC disappeared, wheezing, in the darkness of his personal *BRDM*, and the driver floored the gas pedal.

He spent less than five minutes with me; how much dose would actually be written down for him, nobody knows. My "commandos" were sarcastically smiling, watching this circus from *otsidka*.

Honestly, I didn't care. All of us had our own place in this incredibly complex body of ChNPP.

I decided not to drag the team to *ABK-2* for a shower, but instead to save time and quickly return to the Brigade, catching up with the "bathing day". Just like all squads that worked at the Station, my team had a luxury — if one could call it that — of cleansing after the jumps at the *ABK-2* complex. However, our fellows from the Reconnaissance Battalion, and others in the Brigade, had no such opportunity, and for them a typical Army field bathing facility was assembled at the Brigade outskirts. That day was the bathing day for our battalion and for two of our neighboring units, First and Second Battalions of Special Operations. Due to the specifics of our return trips to the Brigade (the checkpoints, the *PuSOs* and so on), the way back sometimes took several hours and was so exhausting for the teams that the advantage of the *ABK* showering was actually wiped out; by the time of arrival, the soldiers were sweaty and steamy again.

Brigade "bathing facility" was a primitive shower with about thirty or so heads, arranged in a huge tent without the ceiling. Anyone who had ever used the Soviet Army field bath, would understand me, recalling the vast puddles everywhere, the slimy wooden gratings that floated in the puddles and sank deep once you stepped on them, the swamp smell, the lukewarm (at best) water barely dripping from the showerheads ...

Just like at "The Roof" jumps, the bathing day wisdom was tough: to shower more or less humanly, it was necessary to get into facility in the first thirty minutes after its opening. Soon after, the puddles in the shower room were competing with the Pacific Ocean in size, and the "hot" water was barely warm.

Once I realized that we could beat the traffic on the way back and actually — if we played it smart! — return on time to get in the showers early, I shared my thoughts with the team.

The option to "skip the *ABK*" was approved unanimously.

On the way back, we took a small detour: the guys were wondering if the construction of a "sarcophagus" over the destroyed reactor had already commenced. I decided to give them such a chance, since not all from this team saw the "Four" before. Obviously, I was not going to give them a tour, we just stopped on the circle road and looked from afar.

The colossal appeal of two huge radio-controlled cranes, placed on the sides of the reactor building, was almost intimidating. Brand new, shining with yellow paint, on wide tracks. Gorin said that each of those cranes costed over fifteen million dollars; I thought the price tag was grossly understated. Their arms easily reached the bottom of the deaerator stack. They would lay the roof framing and the top panels of the sarcophagus, and would assist in erecting the formwork for casting concrete into the walls.

Somewhat confused, I looked at the *IMRs* and other heavy machinery that were lined up in a single file at the base of "Four", along its front wall. They were parked tightly, head to tail, without gaps. Then I saw the thick, tall metal sheets that were welded snuggly to the sides of several machines in the beginning of the convoy, forming a sandwich; I realized that this would be the bottom part of the formwork. Machines were parked at the very bottom of a future sarcophagus wall, reinforcing the lower part.

There was something ominously mesmerizing in this last parade of heavy equipment irradiated far over the limit, that were spent and not needed anymore. In every next *hodka*, when my route was going near the slowly growing sarcophagus wall, I watched, with mixed sensations, how the "Four" was disappearing from view. While taking these final looks at the crippled and now dormant reactor, I experienced a somewhat disturbing feeling of regret.

All comes to an end.

If I remember correctly, the last time when I saw a grime on the walls of the destroyed reactor building was closer to the end of August. Soon, the wreckage completely disappeared behind the neat new wall of the sarcophagus.

A week later, my team of fifteen was totally worn out, removing abandoned vehicles from the Industrial Zone for the entire shift. Several ambulance minivans, trucks, and even a large bus joined the "Ural" truck that I once used. We loaded them with a mobile crane on a trailer, hauling them one by one to a huge *mogilnik*. There, rusting under a steady autumn rain, hundreds of other cars and trucks, from fire engines and ambulances to dump trucks, buses, cranes, mixers, were parked for eternity. I still have goosebumps seeing the photos of equipment *mogilniks*. Absolutely dreadful. If you want to know how the apocalypse feels, check out those pictures.

With that shift and with the following few raids of garbage and rubble removal, the clean-up phase of the Industrial Zone, the most unrewarding and dirty due to the constant threat of the secondary irradiation, was completed.

On the way back, we decided to cut through the woods to bypass the stanch *Moldovan* checkpoint. However, all detours known to my driver were brutally trenched off by the military police. A never-ending battle between the rogue drivers and law enforcement was expressed in the Exclusion Zone by a staggering number of detours heading from the highway to the gaps in the woods, which were cut off by the trencher and thus were not functioning anymore. Yet the new attempts to go around the omnipresent *Moldovan* checkpoint happened daily, detectable by fresh tire tracks off the highway into the woods; but in next morning, all of them would be cut off as well. Experienced military drivers knew them all, the old and the new; word-of-mouth was the best medium among drivers at the Station. I knew that it was wrong, and by rule we should go through the checkpoint at the Exclusion Zone exit. However, that rule was a nuisance in relation to our Brigade vehicles — we should not be treated as the rest of the vehicles leaving the Zone, because we were not going further than Oranoe village. The solution could have been simple and it had to be implemented way before the Brigade was stationed in the mere 3–5 kilometers outside the Zone. Either our carpark had to be placed inside the border, and Brigade personnel had to be then driven there using clean *borts*-shuttles, or it would have been more logical to station the entire Brigade within the Exclusion Zone. That way, we would never have the exhausting, several hours long, return trips from the Station.

The driver cursed quietly. It appeared that all recent detours were cut off that morning! *Moldovans* were already looming on the horizon, when, desperate, the driver turned sharply to the right, almost flying over a freshly dug trench. I heard a choir of swearing in the back, but we were already jumping over the bumps of the dirt road, rolling into the forest.

"Was there the MP patrol?" I asked. I noticed a small all-terrain car behind, when we were gliding over the trench.

"I don't think so," he replied.

"Kill the engine, I'll check how the gang is doing over there," I commanded. Once he stopped, I headed to the back.

The soldiers griped, but they understood that there was no other way for us to get to the Brigade soon. I gave them five minutes to have a smoke and for "vegetation irrigation". Gorin turned on the *depeshka*. It was clean there, despite the abundance of coniferous trees, which accumulated radionuclides more readily than the leafy ones.

I stood near the *bort* and smoked. It was eerily quiet. Once again, I was humbled by the glory and radiance of Polesye's nature, which was flourishing despite the severe contamination.

Gorin came near me.

"There, look, under the forked spruce," he whispered.

I carefully turned my head and saw a magnificent, huge moose. Just seeing that creature, for the first time in my life in the wild, gave me goosebumps. Gorin startled the animal, trying to drive it towards the Zone border — hopefully, it could escape.

In that summer of 1986, we were absolutely sure that all life in the Exclusion Zone was doomed.

Luckily, we were mistaken, but this is a totally different subject.

We managed to arrive at the Brigade on time, right at the opening of the bathing facility. After thorough showering, I wore brand-new underwear (with tags) and Genyk's gift, a chic black tankman jumpsuit. Mellowed, I sat down in our tent on the bed and pulled out my most craved stash, a three-quarter full bottle of "Pshenichnaya" vodka. The soldiers had brought me a field ration from the canteen: half a loaf of black bread, a can of spam, a large piece of cheese, a pack of cookies "Chess" and a can of condensed milk. Under the bed I kept a case of sparkling water "Borjomi": the miners regularly provided me with mineral water. By the way, a week ago all battalion officers were segregated to a separate officer's tent. Genyk and I refused. What difference did it make where to spend the night, especially if you slept sporadically (within my last few jumps there were the second and the third shifts, late night and wee hours); besides, we were attached to our loudly snoring miners' gang. The commanders did not insist, and the miners, in my opinion, respected us even more.

I took two shots of vodka, ate a little, drank more, then had a cigarette. Then took two more shots.

I wasn't able to get drunk. Not even a little. I looked at myself in the mirror. Looking back at me was a withered, dry-as-a-bone man with a flour-white ghoulish face, black circles under the eyes, a dreary-thin neck and a short bob of greyish hair. My right eye was inflamed, with lids red and swollen. On the neck and on the forehead, in places not covered with a hat, collar or respirator, I had coin-sized pigment spots. Only the mustache shape somewhat resembled someone I knew in the not-so-distant past, and yet, thanks to a strange reaction of the hair with either the radiation or with the respirator chemicals (or both?), the mustache acquired a frightening reddish-orange tone.

The guy in the mirror looked unreal, as if from another dimension. Speechless, I sat down, then sardonically chuckled: I hardly remembered the name of our Department Head. The dimensions of my universe now shrunk to the borders of the Exclusion Zone and territory of the Brigade.

Absolute collapse. I pointlessly stared at the tent wall for a long while.

Probably, the alcohol finally got me. I did not notice how I fell into a heavy, dreamless oblivion.

...Svyat spent a good five minutes waking me up next morning. He said that we must take photos for the pass to Chernobyl — *COS* Zelenov's order. The better deal would have been to get the pass "VSYUDU" ("Everywhere"), which was most valued among liquidators, but it was an unattainable goal; "VSYUDU" gave you access literally to any place at the Station, and *OGMO* wouldn't shower reservists with these.

"Why do I need it, I am not visiting Chernobyl at all," I reasoned languidly.

"Well, think about it as a memory; we all want to get these. Petrov takes photos, I will then go to Chernobyl to process the film, print pictures, get passes and laminate them. Once in a blue moon you could stop at Chernobyl *OGMO*; at least you will eat like a human being," Svyat said, casting a sorrowful glance at the remnants of my feast on the nightstand. "When did you eat your last hot meal?"

I did not remember. "Maybe a week ago? Two?"

"Are you kidding? Wanna get an ulcer, like me?!" Svyat went ballistic.

"Yep. You tell *COS*, that I'd like to get his approval for a pass to the red-light district. In Pripyat," I sassed back.

We schlepped to the Command Tent. Petrov took a picture of me outside, using the tent as a background. The next day Svyat brought me a green rectangle with burgundy letters "CHERNOBYL" on top and my picture under laminate film. I now keep this pass framed, in my office. In that picture, almost fully faded because of radiation, I looked worse than I look now, thirty years older.

... But the main thing was that evening, when I thought with absolute apathy: "It cannot be worse than now," I was wrong.

Three more days and nights passed, during which I jumped at the Station three more times, shoveling out the last radioactive manure from the ChNPP.

As far as I remember, on August 18th we cleaned up the *ORU* 330, a complex system of wires and transformers, supplying the electricity to the high voltage power lines. That was the place where the "Four" spewed a significant amount of the reactor's nuclear filling.

The next day my team had wasted almost the full 8-hour-long shift, blocking and then unblocking the entrance to the courtyard of the Industrial Zone from the side of the Administration building. The job was completely asinine due to the fact that every five minutes after we installed the heavy concrete roadblocks, barricading the entrance, there appeared a car, or a *BRDM*, or a truck, which "by all means and right now" had to pass through that very entrance, instead of going through a second entrance, at the other end of the reactor building. In one of those cars I saw Academician Legasov. Each time we had to use the truck crane to remove two blocks from the road, let the vehicle pass and put the blocks back.

After the fifth or sixth time, the Colonel with *OGMO* badge came running to us, apparently watching our exercises from the window of the Administration building; he screamed with eyes popped out that "you have to shut it down once and for all!" I waited quietly until he stopped spitting saliva, and marveled at my own nonchalance. I "yessir'ed" him and ordered the barricade blocks back. A few minutes after he was gone, a civilian came out of the Administration building; he had a sad, tired intelligent face and asked me very politely, at whose disposal we set up a barricade. I referred to the *OGMO*. He said: "Yes, of course," and went back. Soon after, the same Colonel galloped to us again, now red-faced from humiliation. He told us to remove all blocks from the road and then hit the showers. What exactly did not connect out there with that roadblock remained a mystery, but such mishaps during my time at the Station happened quite often.

Finally, on August 20th I led a team of thirty soldiers to the third shift jump that started at midnight. We combed the Industrial Zone, once again, picking up the last remaining rubbish, which could still be loaded on a dump truck by hand. In the scarce night lights of the Station's backyard I came across something that occupied a special place in my memories of Chernobyl.

Somewhere in the area of the "Two", I suddenly stumbled over a greenhouse. That was an unusual find, and, curious, I wanted to check out what was in it. It turned out that the prudent managers of the ChNPP had grown there *roses* before the war! After the accident, they were abandoned, and without care and watering, the roses quietly died. They stood as they were, when I found them, with proudly raised yellow buds, which were never destined to blossom, resting on yellow stems covered with yellow leaves and yellow thorns. Chlorophyll in the roses was killed by radiation. There was not a pinch of green color in them.

Stunned by the symbolism of the place, I stood at the greenhouse, petrified, for a while.

I now regret a lot that I did not have a camera with me those days.

In the wee hours after that jump, returning to the Brigade, I came to submit a report to the Command tent. Deputy *COS* Petrov was sitting outside. He lost a lot of weight in two weeks of my daily dashing to the Station. I sat next to him. We smoked three cigarettes one after another. There was no talk, but we did not need it.

I did not want to sleep. I did not want to eat. In the full disconnect with reality, I indifferently noted the fact that my last meal was that field ration with vodka.

It was almost three days ago.

The observation did not impress me whatsoever, it was just a fact, that's all.

I knew that in an hour or so Zelenov would again send me with the team to the Station, but I could care less. I had got used to the Station, I knew it to the last brick and door knob. I was its perfect tool, just like its glorious nuclear reactors.

No emotions. No fatigue. Just a full docility and a dumb readiness: if I were ordered the next day to clean up "The Roof" with my tongue, I would not be surprised. I would crawl over there and do it. Stop, that's not it. I would do it gladly, in the most effective way.

I had become an absolute tool of the ChNPP operation.

Everbody was gone. I sat alone at the Command tent through sunrise, mechanically smoking one cigarette after another. *Why sleep? Anyway, in an hour I will ride to the Station again.*

The BC Golubev saw me first in the morning. Probably, I looked quite bad, because even in his callous Army-driven heart, as rough as shark skin, my look touched something.

"You aren't sick, are you?" He asked with his typical tuba-like deep voice. "No? Alright. Listen, after breakfast, you take two soldiers from Zelenov and go to the outpost at Frunzinovka pumping station. You all get the field ration for three days, and don't you dare to show your lousy behind here all the way to the twenty-two hundred on August twenty third!" He growled and walked out.

I could not believe my ears. The sendoff to the pumping station was a Pope's blessing, an exquisite treat, a resort. Svyat told me about that some time back, saying that it was perhaps the best that could happen to us in terms of having a quality rest. Formally, the officer and two soldiers had to ensure the safety of the artesian pump (from where the whole Brigade was getting potable water), heroically protecting it from the terrorist acts pursued by the agents of imperialism. In reality, it was just a sacred time for the three lucky bastards to eat, sleep and just hang around, doing nothing. Teterev River was right there, and savvy habitants of the outpost had kept the fishing rods under the lower bunkbed.

I went to talk to Svyat, experiencing some sort of a... disappointment? Like a child that was punished, I was deprived of my favorite toy, without which the whole universe was torn apart. It was strange to realize how much I was attached to the Station, as if they could do nothing without me there.

"It's simple, man," Svyat remarked philosophically, in response to the news about a miraculous yet suspicious mercy of the BC. "Multiple replacements arrived at night, including four new officers. You are not the Station drainage plug anymore. It is over. Put your Winchester down and let go of the horse, cowboy!"

August 27, 1986
Chernobyl NPP, nineteenth hodka

> *"Bloodwork from our Battalion personnel*
> *was drawn by physicians of the military*
> *hospital more and more often. The previous*
> *analyses data were "all good" per medics,*
> *which was astounding, considering persistent*
> *and similar for all of us symptoms:*
> *headache, sore throat, watery eyes, constant*
> *weakness, radiation burns on the skin."*
> (V.A. Gudov, 731 Special Battalion. ISBN
> 978-966-439-166-2)

I felt like I was a walking urban encyclopedia.

I was graced with an enormous amount of the Station's common folk wisdom. Being armed with it to the teeth, I shared the knowledge with others, often mentoring the "young ones" (If I had a chance to watch "Star Wars" before Chernobyl, I would definitely have compared myself with Master Yoda, but this movie was not shown in the USSR).

That day we jumped to "The Roof" again. My sixth time.

I had a team of twenty troops and a *stager*, an intern officer. He was not my direct replacement. I did not know what went wrong in the exceptionally cumbersome mechanism of military enlistment offices, but the replacement of officers happened extremely irregularly at that time, which explained my Station marathon for fifteen days in a row. There was also a chance that people, frightened by the increasing flow of negative information, began to deter enlistment more and more, or maybe all those who were easy to catch, already went through the Chernobyl meat grinder, and the MPs were hunting the renegades?

The *stager's* name was Denis. He came from Dnepr, too; two years ago, he had graduated from the same Institute of Chemical Technology as I did. This was enough to bestow my warm patronage on him.

Razvod near the Station had been canceled a while back, thus, immediately after a short briefing at the Brigade we headed directly to the Station. Mindful of the lessons of previous jumps to "The Roof", I wanted to bring the team to *ABK-2* as early as possible. We were rolling fast; I did not know the driver, but I didn't have to spur him after delivering a short and sweet message: if we reach the station before 07:30, we would all be most definitely done by 10:00.

I took a peek at Denis from time to time, and his reaction reminded me of Alex Holodov and myself. Denis was very nervous, eyes wide open, eagerly absorbing the scene outside.

The highway to the ChNPP had changed a lot during those weeks. Villages on the roadside were more and more dilapidated without supervision. The foliage on the trees was still feverishly green, but the approaching autumn was imminent. The cereal crops were not harvested and were beaten down by frequent rains. Puddles did not dry out on the roadside anymore, and the white foam on their edges indicated the frequent use of deactivating chemicals.

"What's this?" Denis pointed to the sparkling-bright web of wire antennas hanging on tall thin supports; they could be seen from afar, that's how tall and wide they were.

"Chernobyl-2, or *"Duga"*, top-secret military anti-missile defense system," I replied. "We only hear rumors about it. Seems like it's already shut down. At night, with flashing lights, it looks extraterrestrial, you will see."

In the Station's central square, as always, there was a lot of morning commotion. I recalled that a mere ten days ago there was an impassable swamp, dirt up to the knees. Now it was clean as a whistle; maybe somebody from Moscow, or even from the IAEA, was expected to visit the ChNPP.

We jumped out of the car and ran to *ABK-2*. I did not have to give the team a lengthy explanation on why we had to rush, particularly since I saw three or four familiar faces among them.

The appearance of the Station's Industrial Zone had also changed a lot since my first jump. The territory was clean, the rubbish was removed, the area was coated with concrete slabs, all the way to the very wall of the reactor building. The chaos and the stress of the first months, still acutely felt during my first jumps, was replaced by planned, orderly work on the preparation for launching of the first two blocks. The Station personnel had returned, they were replacing the liquidators. (Somewhere I have heard that in the first two days after the accident over 4,000 of the ChNPP staff, which totals 5,000, had run away, but I don't buy that. Russians are not that cowardly.) They said that the launch was dedicated, as always, to the important historic date, November 7.[3] I didn't know how real that push was, but we all did everything that was required to make it happen, including the cleanup of "The Roof".

I had been there five times, not counting the current jump. An accurate and continuous individual dosimetry for all liquidators in the summer of 1986 was still an unattainable goal (I return to this issue a bit later), and we were not allowed to hold our personal dose accumulation cards until the very last moment of replacement, when we were leaving the Brigade. Therefore, the dreary thoughts of overexposure had always crept into our minds. Of course, no one stated plainly that it was totally safe to get 25 roentgens, and anything above that number meant trouble. Of course, my *real* absorbed radiation dose wasn't determined — and I am dead certain that there were dozens, if not hundreds of thousands of liquidators like me, whose actual radiation exposure was only a guess, and in many cases, like mine, the actual dose was exceeding the threshold of 25 R *four or five times more*. And most certainly, matching my six jumps on "The Roof" with 10–15 shifts spent in the radioactive dust of the Industrial Zone cleanup is totally inappropriate, because comparing the total absorbed radiation dose and the amount of secondary irradiation coming from the absorbed radionuclides is meaningless.

This time I was not going to work on "The Roof". I wanted initially, but then I talked to Svyat. We both agreed that it would not make much sense. *COS* Zelenov also said: "Listen, you're a hair away from the maximum, I do not have any more *coaches*, and newbies are arriving in bunches. So, cool it, alright?"

That's when I felt like Master Yoda.

In terms of the accurate dosimetry records for liquidator personnel in 1986, it is still hard for me to understand as to why through the end of August 1986 the individual dosimeters for liquidators, at least in our Brigade, were limited only to the military sets of DKP-50, fifty

---

[3] One of the two most celebrated State holidays in the USSR.

individual dosimeters with one measuring device. They were made to measure the cumulative radiation dose at the levels between 2 and 50 roentgens. Those who supplied DKP-50 to us should have been fully aware of this feature and should have known that this design made the accumulation of smaller doses using DKP-50 technically impossible; so, for example, the doses that we acquired during the Industrial Zone cleanup, were undetectable and were gone down the drain. We complained persistently, and eventually these dosimeters were discarded. Thus, we returned to the old *depeshka*. After that, the pen-sized aluminium cylinders of DKP-50 were the most common souvenirs among liquidators in August 1986.

Soon after, many new types of dosimeters that were feebly manufactured, began to appear in our Brigade, mostly of the "blind" type, that would not let the owner instantly check the accumulated dose, but he would have to wait until the readout taken by the appropriate device back in the Brigade. The reason was obvious: if the liquidator could see the *current readout* of the individual dosimeter, he could panic and run away from the spot of high radiation. I personally see this as an insult to all the liquidators: with only a few exceptions, we all were committed to work dead on, literally, regardless of the doses and the outcome of irradiation, so whether we saw the real data or not, we would've kept going.

This was the time when I saw an incredible number of new dosimeters on our personnel in the Brigade. They were based on different principles, but all had one drawback: the accuracy of the measurements was not guaranteed. Later on, I read the memoirs of my fellow liquidators, and many of them confirmed the same thought that I had for all those years: the actual doses that we acquired were *way above* those recorded in our military documents, oftentimes as high as 7 to 10 times over the sacred number of 25 roentgens.

The first time I saw those new "toys" was somewhere in mid-August. I went to the wash basin before dinner, and shared the tap next to me with the Captain-liquidator. His trousers and jacket were decorated with a weird collection of plastic and aluminum dosimeters of various shapes and sizes, about a half-dozen pieces. I asked him why he needed all of those. He glanced sullenly at me.

"Haven't you heard of *the rabbits*?" He said bitterly.

"*Rabbits*" in the liquidators' slang meant two comparative groups, each of them containing equal number of liquidators, either twenty or fifty, who were sent to the same places at the Station at the same time. However, *only one group* would receive the radiation-deactivating medicines, the other would not. I mean, both groups would get the pills, but one would be a placebo, and the other the real treatment. *The kicker was that they did not know who received the placebo.*

Working as a pharmaceutical scientist later in my life, I felt that this was perhaps the most callous and demeaning part of all that I have learned at the ChNPP. The *rabbits* were more like the "rats" in the clinical trials, only they were called more decently. The whole purpose was a comparative clinical trial, as they were fed the experimental anti-radiation drugs, one group only that is, but the outcome of this trial was deadly. The *rabbits* were used at the most dangerous places around the Station, primarily on the high fields. After a few weeks they disappeared from the Brigade.

How exactly that "trial" ended, I do not know until this day. I was unable to find any records about the *rabbits* anywhere on the internet. I would very much like to know how many of them survived, however rude and cynical it sounds.

From that time, the "dosimeter rush" had launched in the Brigade. Everyone was trying to collect as many of various dosimeters as he can. This epidemic was further enforced by the fact that our civilian colleagues from the Ministry of Energy and the Kurchatov Institute were generously supplied with all kinds of new gizmos. Dosimeters were exchanged for newer dosimeters, for vodka, for cigarettes … I still kept a DKP-50 cylinder on me, which was stuck somewhere around 2 R (even after my jump to "The Roof" in the middle of August, the dosimeter "hair" indicator did not move an iota!), and another "blind" dosimeter; I do not remember the brand, and I have to confess that I wore it for coolness only, because there was no readout device for dosimeters of that kind in our battalion.

But I digress again.

To my disappointment, a few dozen troops had already landed at *ABK-2* before us. Getting even in the first hundred would not be bad at all, but I would like a more advanced position, so I quietly ordered the soldiers to change on the second floor of *ABK-2*.

I had an additional card up my sleeve, a little-known passage to the "Four" through the "Golden Corridor", a throughway at mark 9 m, connecting all four reactor buildings with *ABK-2*, which was supposedly closed to the common folks from the first weeks of the accident. A civilian customer told me about its existence during either the third or the fourth jump on "The Roof".

A few minutes later we quickly moved along the hallways towards the "Three", and then through the "Golden Corridor" in the body of the reactor building we quickly reached a familiar staircase leading to "The Roof". Running through the "Golden Corridor", I did not experience anything out of the ordinary, based on my perception of the high radiation fields, which was a good thing. With so much radiation coming from the direction of the "Four" through the crippled protection wall, carrying potent energy charge, one could only guess how many dreadful gamma-ray beams we went through; if you crossed its path, you might get a whopping dose.

When we reached the cordon at the end of the corridor, no guards were there yet, and we moved inside the "Three" freely.

The sound of our boots was echoing in the staircase. Level after level, we climbed up, still not meeting the end of the queue for the jump. I was happy! Denis, whom I instructed the same way as Alex once told me — "Follow me step by step!" — dutifully climbed behind me. The rest of the team followed in the rear.

We came almost to the very top of the stairs; ahead I saw the backs of not more than two dozen soldiers. Excellent. If everything goes as usual, we would be out of there in less than an hour.

However, in full compliance with Murphy's law, everything turned out differently. A deafeningly loud sound of Mi-26 blades, coming from the area of the deaerator stack, was a sign of concern: the freight chopper was dispensing the decontamination solution. We had to wait longer, until the levels decreased; it was a very strange phenomenon related to the Station's ever-changing radiation field pattern: every time the decontamination solution stroke the open reactor, the radiation levels around it, and especially on "The Roof", spiked sharply, and then slowly went down. The same thing happened when hard rain hit the Station. I am still unsure what caused that weird reaction.

Anyway, we came to the attic leading onto "The Roof" in only an hour or so.

I helped Denis to get into the "Roof Armor", to put on the heavy respirator. Things had changed quite a bit since my first jump here. There was a lead-filled rubber apron with locks, lighter and obviously more flexible than two lead plates tied with wire; we also had thinner, lighter *propitka*, better boots and gloves. But it did not change the intensity, the stress, the adrenaline rush. Saying something meaningful to a person in this state is useless, I remembered that on my own. One of my soldiers quite cleverly compared a wait during these few last minutes before the jump with the anticipation of guillotining. That morning would be remembered by Denis for a long time. Still, I tried to get through to him.

"Remember every word of the briefing as a prayer. You can forget your name, but all that was said, you have to weld in your memory, because it is your recipe to get back in one piece. Don't you dare get lost on the roof maze, there is no trial and error here!"

He nodded like a bobblehead figurine, but I doubted very much that he had heard all what I said.

Clumsily moving in uncomfortable, heavy protective gear, he jumped out onto the roof. I started the timer. On that day, the levels were forgiving: the shift was up to three minutes long. Two civilians in the attic quietly chatted. I overheard: "... rush is unbearable, I want to peel the skin off ...", "... in Pripyat, they do not let anyone take even the documents left behind, forget my brand-new fridge ..."

Seconds were stretched even for me; I imagined how Denis must be feeling at the moment.

He finally got back. Pulling the respirator down, Denis convulsively gulped the air with his mouth. That was also familiar. Next went the first of our twenty troops. I put in the beginning of our line those three who had already worked on "The Roof". The face of the first soldier did not express much, but sweat marks on the uniform and perspiration on the forehead talked on their own.

The first soldier jumped onto the roof. I continued my "Yoda-go" coaching while heading down with Denis to show him how to change and to shower, back at *ABK-2*. On our way, we reminded our last soldier (I put the team *dozer* at the end on purpose, he seemed to be balanced and mature) about the meeting time and place when the shift was over. The stairway was already full with dozens and dozens of soldiers. Many looked at us with envy.

"How is it out *there*, guys?" Someone asked.

"Cool, man; the main thing is not to psyche, breathe through your nose and don't pee in your pants ... much," replied Denis firmly.

I hid a smile.

When I showered, I suddenly thought that I would never see "The Roof" again in my life. Except perhaps in the nightmares. Recognition of this fact brought nothing; not a satisfaction nor a relief. It was just a statement.

September 1 – September 8, 1986
Oranoe village, Camp of the 25th Brigade of Chemical Defense

"In accordance with resolutions of the USSR Government, a substantial number of the USSR armed force members is involved in the liquidation of the aftermath at Chernobyl Nuclear Power Plant accident. Currently, the total number reaches over 31,000 troops (both active and reserve), located in 43 military units. However, based on our intelligence, these forces are used ineffectively. For example, in early August 1986, only 14,620 out of over 31,000 members were actively involved in the works, which is only about 46% of total workforce available."
(From the Memorandum of KGB Lieutenant General S.N. Mukha to KGB Chief, Army General V.M. Chebrikov; published in a set of declassified KGB documents related to the ChNPP accident, site http://rusnsn.info/analitika/rassekrechenny-e-dokumenty-kgb-po-chernoby-lyu.html)

My final, twenty-third, jump did not leave a long-lasting imprint in my memory. It was August 31st; I remember only that we worked somewhere inside, washing the walls and the ceiling with deactivating solution. On the way back, our super-confident driver decided to bypass the *Moldovans* to the left, through Kamenka and then Gornostaypol, but we got lost and wandered through a maze of dirt roads for a long time, hiding from the MP patrols and unsuccessfully trying to ask the directions to Oranoe from the rare oncoming drivers, who confused us even more. We ended up in Opachichi, very far from Oranoe, but at least then we knew our way back. However, we were too late for dinner, and the disgruntled team harangued the driver and me as well, just for the sake of it.

Maybe I do not remember that *hodka* well because I miscalculated my total dose, and thus was absolutely certain that I would be sent to the Station again, so it *"was just another day in paradise"*.

But the next day I was "benched". Two days later, Genyk was relieved from his daily trips to the Station laundry. Denis, my stager, took over his duty of the Laundry Commander, as we called Genyk.

We became the Station's waste material, and should have been sent back home, but ...

There were two factors which grossly affected the purgatory state of liquidators like us. The incredibly unfair and harsh system of enlisting reservists to the "special training camp exercises for a period of up to six months" and especially the most cruel but justifiable (from the Soviet Army point of view) equation of "25 R or 6 months" gave birth to the expression "to sit on the *fon*" in the liquidators' slang. I will explain this in more details because it was perhaps the most aggravating issue of every single liquidator-reservist with the Ministry of Defense.

The Army wanted a complete accountability of its troops. One guy out, one in. That was simple. No headaches associated with irregularities in the headcount. It was based on soul to soul exchange. Purgatory, like I said. The second factor was tightly depending on this replacement principle, although instead it had to define it. I already mentioned that the cumulative dose of 25 R was a revered number, a winning lottery ticket for every guy in the Army uniform in Chernobyl of 1986. However, these two factors — reaching the 25 R dose, and getting *zamena* — rarely coincided. Most often, if not consistently, there was a time lag ranging from several days to weeks for the unlucky ones. Since every case of personnel reaching 25 R and not leaving the special training camp had to be immediately reported to KGB, the crafty recordkeeping of our doses was issuing everyone approximately 24.5 R by the books (this number was based on the reports of team commanders, and was juggled with at the Brigade HQ). Once this magic number was reached, the poor fella who got it had to "sit on the *fon*", receiving the puny 0.03 R for each day of being stuck in the Brigade — this was the value of the *fon*, the daily baseline radiation level accepted for our area (near Oranoe village). This extravaganza of manipulative recordkeeping continued until your very last day, when your long-awaited *zamena* finally arrived. That day, you miraculously got the remaining roentgens (no matter how much was left, normally these were the odd numbers like 0.46 R or 0.11 R) to bump your sum to 25 R precisely, and off you would go. Then, and only then, you would see for the first time your "Card of Cumulative Radiation Dose", which was always rounded up to 25 R.

On the other hand, if your *zamena* was unavailable, or ran away, deserting his duty, it might take a while to find another "soul" of equal value for yours. What were the reasons of individual selection, you may ask? I had no idea, but all of us knew the name of the officer we had to replace (remember Alex Holodov?). This was one of those Army mysteries, usually performed by some recordkeeper dude deep inside of the military bureaucratic intestines.

Thus, Genyk and I were sitting ducks, waiting for our *zamena* to arrive.

It took us a whole eight days. Eight days of depressive, dull, long waiting.

... Genyk and I slowly combed through the pine grove behind the car park. It was spotted with birch and aspen trees; at a couple of meadows, huge oaks ruled the vast space around. The best place for mushroom picking.

Yellow boleti were squashing under our feet. Their mycelia, whirling and weaving, left hundreds and hundreds of picture-perfect mushrooms that densely covered the clearing. The size of *fly agarics*, the infamous red-with-white-spots poisonous mushrooms, in this forest was terrifying; they ominously shone their red traffic lights at the edge of the meadow.

We did not pick the *yellow boleti* — too simple, too many. We hunted for the "white mushrooms", the kings of local forests, in a competition to collect more.

Radiation in small doses (take, for example, our Brigade) has a beneficial effect on the growth and quality of mushrooms. And of course, I had never seen so many mushrooms in such a small territory in my life.

Polyethylene bags that we carried were quickly filling up with "whites". We collected about a kilogram each, having walked less than a kilometer. It's a pity to throw them away, but it's for the best. Mushrooms are notoriously known for absorbing radionuclides.

However, the locals who still remained in the Exclusion Zone ate them with pleasure. As eagerly as they ate the cyclopean-sized potatoes. The Colorado beetle did not eat radiation-laced potato greens that summer, and the rare farmers in the Exclusion Zone joked that *"dem' bugs know better than people"*; however, they were not squeamish themselves. The locals also milked the cows (which devoured the lush, juicy grass with lots of isotopes in it), and happily caught catfish living in the mudflow of Teterev and Pripyat.

In radioactive silt.

Many locals simply didn't care about *the war* and its toll. However, they thoroughly appreciated the "free" machinery, which could be found in abundance around the Exclusion Zone. It was the peak of the harvest season in Ukraine, and some of the more practical and adventurous farmers even stole the trucks, tractors and cars from poorly guarded parking lots of numerous military camps and even from *mogilniks* around the Zone. In the previous week, we nearly died of laughter, when the battalion Chief Supply Officer told us at breakfast the tale of the final hurrah of the ill-fated *"Mogilnik* Dream", the legendary *bort* MM 00-02.

A week before, it was stolen from the battalion parking lot. The thieves stealthily pushed it by hand through the frail barbed-wire fence (like gypsies, who do wrap the hooves when stealing horses), and successfully got away.

The *bort* was eventually located by the police somewhere in the Mogilev region, in Belorussia.

The thieves should have had a dosimeter, because poor MM 00-02 almost glowed in the dark from the absorbed radiation; naturally, the first checkpoint they went through with it easily caught the vehicle from the Station.

Genyk and I were getting hungry, drooling over top quality fresh mushrooms, but still threw them away, calling it a draw.

We returned to the battalion. Last couple of days we felt depressed. We were useless. We couldn't go to the Station anymore. Plus, our commanders quietly hated us, because by right we were the veterans, "The

Yodas", and assigning us for, say, some stupid humdrum jobs (like that dreadful turf-patching that I had in early August) would be incredibly unfair, and all battalion officers would find this oppressive. That gave Genyk and me an immunity against such chores, except perhaps the work on building the winter flats.

On the other hand, the construction of warmer, sturdier buildings instead of tents was in full swing. Every morning we saw the steady rise of red brick walls at the South-Eastern periphery of the Brigade. If we were marinated in the Brigade for a week or so longer, we would risk getting an assignment over there as foremen. However, at the moment we were just hanging around *"until requested"*, similar to my first day at *perevalka* base in Kiev suburbs, where I was christened into the liquidators. God, it seemed like it was a year ago, but in reality, it was only a bit over a month back.

We anxiously waited for *zamena*. But for the past week, there were only a fistful of those, and only at the level of privates and sergeants. A strong and steady replacement river, which brought me here at the end of July, became quite shallow in August, and in the first week of September it dried up.

Except Deputy *COS* Petrov, Genyk and I were the longest "survivalists" in the battalion. The rest of our "homies", reservist officers, had been replaced. Svyat also left, cordially but abruptly bidding farewell to us: he had to be urgently hospitalized. Most of Kharkov-based officers were gone, and our "Officers Club" was now filled with the new faces. Although the newcomers were not worse, the sense of loneliness seeded by the departure of friends was felt by us deeply. We continued pulling the leg of newbies in the dining room with the classic Svyat's skit:

"Yesterday I was watching a crow flying over the "Four" ... Flapped the wings, flapped, then ... "

"Then what?!"

"Wasn't flapping anymore. Burnt in the air! Poor tiny fireball, fell directly in the crater ... "

By that time, the sarcophagus was already roofed, and while it did not make the joke less current, somehow it wasn't funny anymore.

Genyk's miner team was also gradually replaced, although the orchestra-like snoring mysteriously remained in our tent.

The evenings were now occupied by extended dinners and leisurely conversations in the Command Tent. Additional recreation was accomplished by watching movies on Wednesdays and Fridays. We were shown (in my memory, for the second time) the whole cycle of "Angelica — Marquise of the Angels" (five movies!), and several comedies with Louis de Funes and Pierre Richard; all copies were, as always, in terrible condition, but who would complain in our situation?

One evening in the "smoking room" (four long benches squared around a barrel filled with water at half volume, stale and brown from the floating cigarette butts), I met a young Major, a doctor from the Brigade Medical Company. We talked about the short- and long-term consequences of the accident, specifically the impact of high radiation fields and the secondary radiation on the health of liquidators. From him, I heard a chilling prognosis. First, he briefly mentioned a visit of some clandestine *"medical and statistical group"* formed from the radiation specialists of the Ministry of Health and the Ministry of Defense (he mumbled something of that lingo but I wasn't sure that it was close to what he really meant — he was fairly loaded with alcohol; perhaps, he realized that he spilled the beans, but was too drunk to stop talking). Then he quite casually dropped the bomb by saying that according to the estimates of this group, the long-term prognosis, or rather the lifespan for most liquidators, who were working at the Station or in the high radiation fields in the summer months of 1986, should be limited to 15–20 years from then onward.

That's what he said. Verbatim. And then he hiccupped and puked.

I found his puking in the common room more offensive than the information that he casually shared, and left without saying anything. His words, albeit not completely devastating (just because *I still had those 15–20 years of life ahead*, counting from September, 1986!), were buried deep inside of my mind all these years. Each time I had another health issue (as all of the liquidators of 1986, I had my fair share of these), these words rose from the depths and bugged me — all the way to 2006, the twentieth anniversary, when I took a deep breath and said to myself that if I ever met that guy again, I would break his nose.

However, when I was writing the Russian version of this book in early 2008, I was combing the internet in the search for the real liquidator health forecasts and came across the following lines:

"...*Cases of radiation-induced solid tumors among former liquidators are expected to reach their statistical peak in the near future, specifically in the timespan of about 25 years from the Chernobyl accident.*"

Since then I just stopped paying attention to such information.

But then, in 1986, we did not think so far ahead, although our bloodwork, despite all the secret memos and orders of the Ministry of Health to make it look fine, was stubbornly deteriorating with every jump, and medics were ashamed of hiding the real data from us — well, at least those doctors who truly cared. We did not need to see the results of our bloodwork. We *knew* that we were hitting the rock bottom of our health; when I finally got home, for a while I was unable to walk a hundred meters without sitting down and catching a breath, and I was an avid basketball player, semi-professional in younger years.

I digress yet another time.

... Our last days in the Brigade were all crested on waiting for *zamena*. It was actually scary, how quickly just a week of doing nothing had transformed us from the Station's absolute tools into dispirited, half-sick, half-sleepy human shadows. When I took my weight measurement at the Brigade Medical Company around September 5th, for the first time for my Chernobyl stint, I thought that I misheard the nurse.

I lost about ten kilograms. Ten. For a little over a month.

On September 8th I laid down on my bunk bed after lunch, vacuously staring up at the braid of metal springs of the bunk above me. Genyk entered the tent and said casually:

"Come on, pack your sh*t, *oot*. They are replacing us ... "

I did not express any emotions. None.

We bid farewell to the guys. Petrov returned to us our Officer Identity Cards, with inserted radiation records, from which I was amazed to learn that as of that morning I had acquired a cumulative dose of precisely 25 roentgens, and not a milliroentgen more (!), as well as our travel documents and money. We hugged and hid suspiciously wet eyes. He now remained the only one from the legendary "Old Poopers" squad.

We were in a hurry. The all-terrain car from the Third Battalion of Special Operations was going directly to Kiev. I returned to the tent and took a final look at my place. Like the Pripyat residents, I was taking with me only the bare necessities.

Unlike them, I know that I would not return there, even if I wanted to. I was done.

Genyk and I looked at each other in disbelief: to think of it, in a couple of hours we would be in Kiev!

September 8 – September 9, 1986
Kiev

> *"He who trusts the world, the world betrays*
> *him."*
> (Hazrat Ali Ibn Abu-Talib)

The streets of Kiev appeared to be just the same as I remembered them from *before the war*. The car stopped near the supermarket at the capital outskirts. The officers, who gave us a ride, were going to stuff the car with vodka. We said goodbye and took a tram, heading to the city. It's getting darker. We agreed that we most probably couldn't get to the *perevalka* that day, where our civilian clothes were stored.

We needed to look for a place to crash for the night.

Genyk had no friends or relatives in Kiev. I had two options. The first was my old friend, a roommate from the Academy of Sciences dormitory, who had a one-bedroom apartment on the Soviet Ukraine Avenue; he had always welcomed me at his place during my rare business trips to Kiev. The second was my cousin Peter, the electrician from ChNPP, who upon the evacuation from Pripyat, received an apartment somewhere on Balzac Street, in the area where many of the ChNPP personnel were relocated, and supposedly settled his wife and two sons there. I knew the address; however, I had no idea how to get there. Both places were relatively new and had no telephone, and they were far apart.

We opted to try my friend's place first, which was closer to our drop-off point.

Only after descending into the metro, we understood how stupidly brave was our march across the whole of Kiev in the liquidator uniform. We both were wearing similar paramilitary outfit, but with tops of one design and bottoms of another. We had the same gas mask kit-bags (no masks, though, just our most precious belongings in them; the rest of our stuff was residing in the Brigade-tagged thick polyethylene bag with an alarming yellow label: "CAREFUL! RADIATION HAZARD!"; they soon would find their resting place in one of the *mogilniks*), wore the same camouflaged officers' "pot" hats. From the side, we looked like two serious military dudes with gas mask bags, quite nervously watching the evening crowd.

Around us was a constant human vacuum. We were shunned, all eyes were turned away from us. An honorable distance of two vacant seats on either side of us was kept by other passengers, even though the car was quite full.

The same story happened in the trolleybus.

So far, we did not pay any attention to this vacuum. We were gawking at crowds, nicely dressed in colorful clothing (civilians!), a lot of pretty women, city transport, well-lit streets, shops ... We realized how wildly different was the life there, than in Chernobyl; different in everything — from the colors through the atmosphere to the overall life intensity and its values.

We were way too different now.

My friend was not at home. We waited a little, smoking near his entrance door. We were already quite tired and darn hungry. Down at the entrance to the building, a trio of old ladies, alarmed by our appearance, synchronously moved to the edge of the bench. They, however, pointed us to a restaurant in the nearby mall, two trolleybus stops away.

The restaurant was full. The head waiter, pouting fat lips in disgust while despicably looking at our uniforms, had advised us to take a shower first, then change our wardrobe, and only then to worry about the food. My buddy turned white, clenched fists and started moving toward the buffoon, but I grabbed and pulled Genyk out to the street. There, we took a deep breath and for the first time looked around without the blinders on our eyes.

Kiev, the city that we saved, did not want to embrace us.

We were not welcomed here. We were the outcasts, lepers at best. No one cared, no one actually cared.

Resentment and a sense of huge prejudice hit us simultaneously.

Working at the Station, we naively believed that we put our lives on the line, and it would be fully recognized by our grateful countrymen. Certainly, I did not imagine that our fellow citizens would line up, eager to wash our feet upon returning home, but ... Would there be anyone who would argue that we did not deserve it?

That evening, our illusions vanished.

At best, we were ignored. Most of the times, we were looked at as if we were the carriers of serious infectious disease. Not only our clothing: the uniform, the bags, the hats, — we ourselves were taken as dangerous in the eyes of the people around us. The taxi driver, whom we had to give a *quarter*[4] to give us a ride, shared with us a rundown on that. He himself believed that radiation was not a flu, so he agreed to take us to Balzac Street. By the way, he was the only one out of five or six free taxis who dared to slow down near us, while we were waving in despair at them.

Balzac street was located on the very outskirts of the city. As in all new neighborhoods in the USSR those days, the buildings did not yet have numbers outside. The wind howled in the alleys, chasing the eternal sand dunes, rolling them across wide empty streets — the feature typical for the areas built on the sandy left bank of Dnieper River. It was about eleven at night. A man, walking a German shepherd, pointed to one of the courtyards surrounded by the drab sixteen-story buildings.

Yes, it was the right building. Encouraged, we ran up the several flights of stairs … but no one opened the door.

I realized then that Peter probably lived at Zelenyi Mys Camping at the moment, where the current long-shift Station personnel were positioned (they were commuting from there to the ChNPP), and would stay there for a month before returning to his Kiev flat. His wife Lyudmila and the children, probably, went to their relatives in Berdyansk. Disappointed, we went downstairs, where we held an operational meeting, like in the Zone.

A long-term experience of traveling to Kiev on business told us that looking for a vacancy in Kiev hotels, especially at night, especially on weekdays, was crazy and that we had zero chance. But we had no choice, and since the only other option was to kill the time at the bus station while waiting for the morning bus to *perevalka*, we began to look for any set of wheels moving back, towards the center.

In a dreary cold and damp night, two strange figures in a paramilitary liquidator uniform were slowly walking along the highway towards

---

[4] Twenty-five rubles.

Kiev, hitching rare rides. We were lucky: a *Zhiguli,*[5] owned by a former ChNPP compressor station technician, picked us up and drove straight to the center; the nice man fiercely refused to take any money. His girlfriend, a plump and absolutely drunken blonde, waved good-bye to us from the back window, when we, totally exhausted and cold to the icicles under our noses, stopped at the centrally located hotel "Red Star". The thinking oil in our frozen brains did not flow freely, because the only rationale that we had, while entering the exquisite vestibule of the "Red Star", was primitive: where else would the two employees of the Ministry of Defense go for spending the night, if not at the hotel that was subordinate to the Ministry of Defense?!

Although the bold strike seemed to be doomed to fail from the onset, shockingly, the concept worked. After a few minutes, in total disbelief of our luck, we were already looking out the window of an absolutely gorgeous two-room suite on the Peoples' Friendship Square. The administrator-on-duty took pity on two tired liquidator officers, making us swore that we would check out by seven in the morning. It was close to 2 am. I decided to continue trying our wild luck and completely audaciously called the administrator, asking if we could have some tea in the room. Something had changed up there in the skies: soon, a pretty maid knocked at the door and brought us a pot of the hot sweet black tea and insanely delicious sandwiches with salami and cheese. We scraped off the last Chernobyl mud and dirt under a scolding-hot strong shower, swiftly cleaned up the food plate, gulped the tea and dozed off the moment our heads touched the pristine white pillows.

The last day of my epic was far less entertaining. Genyk and I got to *perevalka*, where the same stifled *prapor* signed off our civilian belongings. Thankfully, rats and mice did not touch them, but in the corner of my bag I found a patch of mold. Wearing jeans, I again realized the fact of my significant weight loss. The belt, buckled at the last hole, barely held the jeans on the protruding hip bones.

The Kiev railway station met us with multiple MP patrols, a huge number of completely different-looking (than us) liquidators, who were

---

[5] A common Soviet passenger car.

viciously hounded by the said patrols, unbelievably large civilian crowds and wild queues at the ticket offices. After some yelling, nagging and waving of our "CHERNOBYL" passes in the air (finally, at least some use!), we bought the tickets: I got SV[6] to Dnepr, and Genyk — some weirdly complicated transfer ticket to his beloved Donbass.

Our last dinner together went well, just as we imagined it should, when we planned it at the Brigade. We scored a table in the restaurant "Khreshchatyk" (allegedly the top food joint in central Kiev), closer to the kitchen, and ordered almost everything that was available on the menu. In combination with our more than plain outfits, the pricey action irrefutably convinced the waitresses that we had recently been freed from the *"zone"*.[7]

We really were from the Zone, but not the one they feared. Much more dangerous.

A bottle of vodka made us more forgiving to the treacherous city.

At some point, we ordered mushrooms in sour cream, baked in clay pots. They were not as grand as ours, near the Brigade, but were absurdly tasty. We left generous tips for the waitresses, who were still wondering about our background, and left for the train station.

Genyk walked me to my train car, since his train was departing later. We embraced each other in a heartfelt goodbye hug. Nothing was said.

I was alone in the compartment. When the train moved, I fell down on the bed, powerless. Something wet and salty was running down my cheeks. Half-awake, half-asleep, I was searching, hearing, breathing, looking, suffocating, sweating, mumbling, yelling, lifting, dropping, washing, running, jumping, scraping, shoveling, stretching... I crouched, shuddering, trying to shake off an intense vision. Some unstoppable, brutal force was dragging me all over the Station. I saw dozens, hundreds of faces, familiar and not, that surrounded me in the giant carousel. I choked with the rotten-sweet air in the Station bunker, I frenziedly chopped the bitumen laced with *TVELs* on "The Roof", and agonized as a butterfly, dying near the withered yellow roses in an abandoned greenhouse of the Zone ...

---

[6] Night sleeper, top quality train car in the USSR.

[7] Prison, Russian slang.

And above all that in my nightmare, spreading its ruined, blackened, gargantuan walls like wings, hundreds of meters up, was the satanic crater of the demolished "Four", relentlessly sending the blue-green glow of fierce radiation high above in the space. A prodigious dragon roar, a non-stop deafening sound of explosion hammered my ears, and I bled through them, as I did through my eyes, until the forbearing nothingness had swallowed me fully.

... The quiet dinging sound of a teaspoon, swirling sugar in the glass somewhere in the neighboring compartment, woke me up.

"Tea? Or coffee?" A car attendant, a young girl in a neat USSR railroads uniform, smiled at me, when I opened the door. A strong smell of good coffee hit the nostrils that hadn't experienced the aroma for so long, and I smiled in response. The first morning of life where I don't have to be concerned about the contamination of food that I eat, liquids that I drink. No need to keep a spare *lepestok* in the side pocket. No more revolting feeling of *propitka* rubbing against my skin. No worries about the doses, levels, dust, *clean* clothing, *razvod*, reports, rule "two-four", *PuSO*, and darn *Moldovans*.

I squinted and cleared my throat, trying to hide an intense, awesome feeling that instantly saturated my heart.

My Chernobyl was over.

RBMK Reactor: General Overview and Main Technological Features

It is hard to expect that over 30 years past the Chernobyl catastrophe many people will remember details of the event, particularly in the unaffected areas of the world. Several of my beta-readers have expressed an interest in knowing more about the background associated with the tragedy at ChNPP. Realizing that there are several books covering the subject, I nonetheless decided to offer three short pieces describing the features and the drawbacks of the RBMK reactors (Appendix A), the factors that, from my prospective, greatly influenced the accident and affected its aftermath (Appendix B), and the brief chronology of the first minutes, hours, and days associated with the accident (Appendix C). These small narratives should help the reader to see a wider canvas of the event that spelled the end of the Soviet Union.

I would like to make clear that the following summary is by far not comprehensive and it just highlights the relevant, from my point of view, technical and historic essentials that are connected with the Chernobyl catastrophe.

The history of the Soviet gambling with the powerful potential of nuclear energy was painted in dark tones. For many years, we lived in oblivion because of the strictly obeyed cover-up of each and every mishap related to nuclear power exploitation. Over ten major accidents during 1965–1986 caused leaks of radioactivity to the environment as a result of improper equipment handling or plain negligence, resulting in explosions, fires, and structural collapses with subsequent overexposure of personnel to radiation. These were never mentioned in the media, except two cases, Armenian NPP and Chernobyl NPP, which both occurred in 1982 during Secretary General Andropov's rein — both were mentioned very briefly in major newspapers. The sad "champion" of multiple accidents within a span of over fifteen (!) years was Beloyarsk NPP, where, due to the frequent misuse of unit 1, the fuel assemblies were continuously deformed and overheated, which resulted in multiple cases of repair personnel overexposure to radiation.

Our people were led to believe by the Party and the government that nuclear power plants were just as safe and harmless as locomotive fireboxes, and nuclear operators were their adorable stokers. Safety protocols and requirements, particularly during the construction of nuclear power plants, were routinely bent and bypassed in order to finish the subsequent construction phase within quite stringent, often unrealistic, time frames. These were in turn linked to certain "milestones", calendar dates, as it was an accepted practice in the Soviet Union in order to receive better bonuses and salary raises, which were quite hefty for the upper management and Party bosses. The history of the Chernobyl plant construction and operation is full of proud reports on completion of multiple tasks "well before scheduled timeline". *This was the all-state custom.* However, it is one thing if you skip or neglect some insignificant phases in building the toy factory (although don't get me wrong, I am appalled by this concept altogether), but completely different for the corner-cutting practice in the construction and operation of the nuclear power plant.

Coming back to the comparison of operators with stokers, I deeply believe that a dilution of experienced staff (caused by the rapid growth of the nuclear power-driven energy sector in the 70s and 80s in the Soviet Union), which eroded the whole body of personnel involved in management, design, development, and operation of nuclear reactors, resulted in the gaps being filled with less skilled but overly enthusiastic people and in the euphoria of seemingly much wider safety margins of Soviet nuclear reactors than it was in reality. This was particularly dangerous in relation to the *RBMK*[1] reactor type — the one that exploded that dreary night in April 1986. This is a class of graphite-moderated nuclear power reactors of high productivity, designed and predominantly operated in the former USSR. Five key features define

---

[1] (Russian: *reactor bolshoi moschnosti kanalnyi*) High power channel-type reactor.

this powerful reactor of high capacity, which was designed using a simple approach allowing to be replicated in large numbers:

(1) channel concept (core consists of a number of so-called technological channels that are positioned in the massive body of the moderator);

(2) heterogeneity (nuclear fuel is separated from the moderator and other elements of the core);

(3) graphite/water tandem (heterogeneous reactor that uses graphite as a moderator and regular (light) water as a heat transfer agent);

(4) boiling type (the steam that spins the turbine is generated directly in the core);

(5) thermal neutrons principle (nuclear reactor uses so-called slow or thermal neutrons).

RBMK reactor's safety was a subject of much debate long before the accident, more so in the aftermath of the catastrophe; however, pre-accident woes and concerns were gagged by a powerful lobby of the designing institutions (Kurchatov Institute and Scientific Research and Construction Institute of Energy) and the upper management of the exploiting organization (All-Union Industrial Department of Nuclear Energy). In the early days of introduction of *RBMK* to the Soviet energy market, it was a firm notion that this reactor type is very safe. I am not going to talk about this historic fib. After the accident, hundreds of corrections were made both in the design and in the safety protocols of *RBMK*; nonetheless, the majority of *RBMK* reactors were eventually decommissioned in Russia and some of the former republics of the USSR. Some say that the popularity of *RBMK* reactors in the USSR nuclear energy was strategic in the sense that due to their high energy and productivity it was possible to use it for military production of plutonium, but (luckily for the world) in the USSR the military grade plutonium was manufactured in reactors similar to *RBMK* with code name PUGR[2] and under far more stringent safety control

---

[2] (Russian: *promyshlennyi uran-graphitovyi reactor*) Industrial uranium–graphite reactor.

and exploitation protocols that were superior to the civilian *RBMK*. They were operated by well-trained and experienced military personnel; therefore, *RBMK* reactors were used for generation of electricity.

At the risk of losing some of the less patient readers, I nonetheless would like to give a crash course on the main features of the *RBMK* reactor. This is necessitated by the need to understand what exactly happened (and how the dynamics of the accident were progressing) during that doomed night. Please bear in mind that the following description is given solely to illustrate *RBMK* key elements and features.

The main part of the *RBMK* is a reactor vessel that consists of the steel cylinder-shaped body, about 14 m in diameter and 10 m in depth, closed from both ends with the upper and the lower biological shields (UBS and LBS), heavy formations aimed to prevent radiation from being disseminated out of the active zone. UBS is a cylindrical disc 14x3 m in size and several hundred (some over a thousand) tons in weight, with a number of vertical holes in this lid to give access to the vessel content — fuel assemblies and control/absorption rods. LBS is designed similarly, only somewhat thinner (about 2 m), and also contains openings for feeding the content into the vessel, but from the bottom. LBS is additionally reinforced by the massive metal plates, which support the whole vessel.

The reactor vessel is surrounded consecutively with several cylindrical bodies serving both protective and technological purposes. The steel chamber is jacketed with a water tank, about 3 m of water thick, split in 16 vertical containments. This water is used as a biological shield and also as a coolant for the vessel. It is followed by a bulky cylinder of sand and finally by the reactor pit, a containment encasing the whole reactor in a thick layer of concrete.

The content of the reactor vessel that produces the energy consists of the moderator core, assembled from separate blocks roughly 25x25x25 cm, which are made from reactor-grade graphite. The use of graphite as the moderator material is aimed to slow down the prompt neutrons, converting them into thermal neutrons. The blocks have multiple vertical channels piercing through the whole height of the graphite core in parallel fashion. The channels are strategically spaced in the moderator body, in a carefully designed pattern to create a steady fission from the individual fuel sources. In these channels, the pressure tubes and the absorber/control rods are positioned. Pressure tubes, each approximately 7 m in length, are the stationary elements of the reactor and are fitted with the fuel assemblies, each containing 18 zircalloy-made thin tubes (fuel rods) filled with the actual nuclear fuel — partially enriched uranium oxide pellets. These rods, *TVELs*, represent the main technological element of the reactor that produces the heat. Two of the fuel assemblies, placed one atop another, occupy one pressure tube. Each tube is individually cooled inside by the running pressurized water which boils in the tube and raises up at the temperature of approximately $300 °$ C. Overheated coolant is directed to the steam separator, where the steam is split from hot water and transferred to the turbine to produce electricity in the generator. The steam is then condensed and fed back into the coolant circulation. Two separate water coolant loops each with four pumps circulate water through the pressure tubes to remove most (about 85%) of the heat from fission; the rest is removed by the control rod channel coolant.

Control rods, or absorption rods, consist of a long graphite-filled section at one end and a boron carbide neutron absorber section at the other end, connected by a telescopic bar. This design reserves a water-filled space in the bottom of the channel tube where the rod is positioned, and in theory allows a smooth control of the fission rate, or neutron flux, in the active zone by absorbing the neutrons with efficiency that is based on the depth of control rods' immersion in the core. There are also some stationary boron-based absorber rods that are strategically positioned throughout the core (their quantity varies depending on the "freshness" of the fuel), but most of them are designed to move up and down in the channels to deliver more precise regulation of the fission process. The movement of control rods is provided by individual servo motors, with the speed of about 0.4 m/s. Lowering or raising some of the control rods will allow the fission to slow down or to increase, therefore cooling down or heating up the core and thus affecting the output of steam generation.

There are 1661 fuel channels and 211 control rod channels in the graphite moderator of 1000 MWt *RBMK* reactors of the second generation — the one that was operated in ChNPP unit 4.

It would be unfair to state that the safety precautions in the pre-Chernobyl design of *RBMK* reactor were marginal. Multiple, quite efficient and advanced for their time, features of hydraulic, mechanical, gaseous, and certainly nuclear components of the *RBMK* reactor were created to avoid potential failures, either individual or concomitant. A vital part of *RBMK* safety was the control and supervision system, allowing the reactor operators to have timely information about the core condition, as well as about various technological processes related to the generation of power. Without going into too much detail, I would like to mention some key features of the *RBMK* safety system that were essential in the accident progression; perhaps the most important was the Emergency Core Cooling System, ECCS,[3] which was designed to control the reactor output if the coolant supply to the core is interrupted; the most frequent cause was considered to be a loss of power. It may sound strange (loss of power in the power-producing plant: an old Russian saying expresses similar odd situations "*the shoemaker who does not own the shoes*"), but with such an enormous source of energy at hand as a nuclear reactor, losing power even for a very short increment of time could impose serious if not catastrophic consequences. ECCS had three separate circuits, integrated in the reactor cooling system, each containing dedicated water tanks, hydraulic accumulators, pumps, and an elaborate network of pipes. When a total loss of power occurs, the ECCS pumps were supposed to be powered by the inertia, the kinetic energy of the running generator turbine blades (as long as the rotor keeps spinning, some electricity would still be produced, until the backup diesel generators begin to work). This attribute of the ECCS, in essence, was a key piece of the system test at ChNPP unit 4 that led to the catastrophe.

It is important to recognize the role of the following four specific characteristics of *RBMK*-1000 in the accident. They may not be easy

---

[3] Russian: *Sistema Avaryinogo Ohlazhdeniya Aktivnoi Zony.*

to grasp for a novice, but are must-know features of a reactor possessing such a monstrous strength and capacity.

(1) *Positive void coefficient (PVC)*. *RBMK* units are cooled by a single loop ("light", regular) boiling water, which produces the steam in the core; its amount varies at different stages of reactor operation. "Light" water is far better coolant and more efficient absorber of neutrons than steam; therefore, a shift in the ratio of steam/water that is called the *void coefficient of reactivity* (with steam existing in the form of bubbles, or "voids") in the favor of either steam or water would result in a drastic change of a fission process. When the void coefficient is positive, an increase in steam amount will lead to the reduced neutron absorption, and this leads to an increase in the reactivity of the system. Normally, the void coefficient is only one of several constituents determining the neutron flux, but in *RBMK* reactors it is the dominant component that highlights a high degree of dependence of the core reactivity on its steam content. When the power began to increase during the accident, more steam was produced, which in turn led to additional surge in power — the deadly loop was created, resulting in a rapid surge in power to around 120 times over the reactor's listed capacity. It has to be noted, however, that increasing positive void is not a catastrophic situation and *under normal operating conditions* it can be managed because of the certain inertia of the nuclear reaction to an increasing number of neutrons (the effect that is called *post-fission neutron emission from daughter nuclei*), but when a reactor is very low in power (which actually happened that April night with unit 4 at ChNPP), the runaway state is quite likely to happen due to a large and poorly controlled positive void.

(2) *Operational Reactivity Margin (ORM)*. Roughly, this is an absolute minimal number of absorber rods that have to remain in the core at different places (strategically calculated and defined by the unit output, spent level of fuel, and the reactivity of the core). For *RBMK*-1000, this number should have been kept at 28–30 [1], although the ChNPP personnel were inclined to

believe that this number was much lower (around 15). The ambiguity of this obviously very important safety feature was in part the result of a massive coverup of each and every — even small — accident in the USSR nuclear power industry not only from the public, but even from the specialists. All materials and facts related to such accidents were classified and were available to only a very limited number of personnel, which, sadly, were not the ones who operated similar nuclear plants and could have (should have) learned from the past mistakes.

(3) *Positive scram effect*. SCRAM, or Safety Cut Rod Axe Man,[4] is in plain terms a mechanism of reactor emergency shutdown. The slang word "scram" defines the need of a swift and urgent departure, and in technical meaning this word is fitting the purpose well. Upon reaching the near-runaway conditions in the core, when other means of controlling the reactor are either ineffective or exhausted, scram action inserts abundant masses of the substance, which is highly efficient in neutron absorption, directly into the core. This action causes the reactor power to drop almost instantaneously. In the *RBMK* reactor safety design, SCRAM is accomplished by pressing the button *AZ-5*,[5] which prompts insertion of all control rods into the active zone. The huge difference of the moderator (graphite) and the neutron absorber materials is that during the fission process the moderator (graphite) is unable to fully stop the newly created prompt neutrons that are piercing the graphite body — it only slows them down and diffuses the produced heat; however, the prompt neutrons are readily caught by absorber atoms, which stops the fission progression. Positive scram effect is an unavoidable design flaw of the *RBMK*-1000 reactors that was not considered as lethal in the pre-Chernobyl era, mainly because before

---

[4] Acronym that states the task of a technician readied to "axe the ropes" of the safety rods of the Chicago Pile One reactor (Manhattan project); in reality, the "axe" action had to be fulfilled by pouring the solution of neutron-absorbing cadmium salt into the reactor.

[5] (Russian: *Avariynaya Zaschita*) Emergency protection.

there were no cases in the history when the combination of factors such as those on the night of the test (see below), that caused catastrophic consequences. This flaw is attributed to the layout of the control rods and their positioning in the core (in the channels filled with water). At their highest retracted position, the control rod graphite sections are positioned at the center of the active zone, with water occupying slightly over 1 m of the channel length from the bottom of the core. When control rods are moving down the channel, water is being displaced with graphite, which is less efficient in the prompt neutron scattering than water. This creates a local surge of reactivity at the bottom of the core while the graphite part of the control rod passes through that section. This is in plain terms the positive scram effect. It was never anticipated before the Chernobyl tragedy that a positive scram effect would lead to such a disproportionately elevated and fast local fission output at the bottom of the active zone, followed by the deadly loop of positive void.

(4) *"Iodine pit", or xenon poisoning.* Xe-135 isotope is formed in the process of uranium radioactive decay. Unlike its predecessor in the beta-decay process, isotope of iodine I-135, Xe-135 is a far better absorber of thermal neutrons with a longer half-life. There is a complex, quite delicate, balance in the uranium decay process, which is now studied in detail, and modern day nuclear reactor operators are acutely aware of the issue related to the buildup of Xe-135 in the core. In theory, if this isotope is produced during decay process in sizable amounts, it absorbs copious quantities of neutrons and thus can potentially restrict fission (this was indeed first observed during the Manhattan Project work, when the researchers found that they had to increase the fuel concentration to overcome the xenon poisoning). With the reactor running at higher energy, fission creates enough neutrons to "burn" xenon atoms, which were produced during the decay process, and the equilibrium in production and burning of Xe-135 is achieved. Increase of energy shifts the equilibrium toward xenon burning, and the reactor gets increased reactivity. When there is not enough energy generated, one of

the ways to bring it back up is to remove some of the control rods from the core to spur the nuclear reaction. However, for *RBMK* this delicate balance between the produced energy output and xenon poisoning of the fuel rods is highly important, which was not taken into account by the operating staff of unit 4.

Chronology of the Accident and the Aftermath

Initially, I thought that it would be a bit irrelevant for the purpose of this book to write about what had actually happened with the unit 4 reactor that night, what was possible to avoid and how the immediate actions of the shift personnel during the accident and the first response measures affected the whole situation. The reason is that there are multiple sources — some credible, some not so much — that describe the timetable and events, which are related to the ChNPP accident. However, it appeared to me that I could perhaps bring up a description of the accident events under a slightly different angle, highlighting events that played, in my opinion, an important role in the history of the Chernobyl catastrophe. It will not be a comprehensive and meticulous description; for a curious reader, there are other books and papers, unfolding the time line in details.

On April 25th, 1986, unit 4 of the ChNPP was prepared for a scheduled maintenance. The plant management had decided to use this opportunity to perform a test of *RBMK*-1000 reactor (the reactor was running at a reduced capacity), targeting the possibility of engaging locally produced power supply (by the unit turbines, specifically, employing turbine rotor *vybeg*,[1] which can continue to rotate for a while even without steam pushing the blades, due to a high inertia of the rotor) rather than using a grid supply for running the reactor systems. It is noteworthy to recall that a similar test was already attempted with unit 4 in 1985, which then failed because of the malfunction of satellite equipment.

At the time of test, the unit 4 reactor core contained over 1,600 fuel assemblies charged with about 200 tons of uranium dioxide, and some 800 tons of graphite.

The test design, which was drawn and approved by the plant Chief Engineer N. Fomin, included a shut-off of the ECCS (Emergency Core Cooling System) to prevent the interference of the safety system with

---

[1] Turbine runoff.

the test progression. Some sources doubt that this decision had a grave impact on the accident; others consider it as a keystone of the series of flaws and mistakes connected to the doomed test. Other safety features also had to be disconnected, which was a direct violation of normal operating conditions of nuclear reactor. However, the ill-fated experiment was designed and scheduled by the specialists in power production, without required nuclear safety measures to be observed and implemented in the protocol. As such, the unit 4 shift operators involved in the test were dubious about the potential nuclear reactor dangers related to the power adjustments during the test. They were also unclear as to how the altered measures of reactor safe operation could affect its stability and power production by the unstable reactor.

It is also obvious that Fomin, together with his Deputy A. Dyatlov and ChNPP Director V. Bryukhanov, did not seek approval of test protocol with respective nuclear safety overseeing organizations, and took the authorization of the protocol upon themselves. No nuclear safety specialist in a sober state of mind would have allowed the test to be performed with *all safeguarding features* shut off. And yet it was done.

Reactor power decrease had begun as scheduled, at 1 am on April 25th. The ECCS (according to protocol) was turned off at 2 pm on April 25th, and by crucial coincidence, exactly at 2 pm, the dispatcher from *Kievenergo*[2] had requested to keep the unit running at 50% capacity to cover the power deficit in the grid. The test was thus postponed, since the reactor power decrease had to be stopped.

Similar tests, with rotor inertia-produced local power, were done before, on other *RBMK* reactors, but only with the reactor cooled down, with SCRAM system fully engaged. The altered protocol for unit 4, instead of bringing the reactor to a shutdown, meant to decrease the reactor power to 700 MWt and then begin the test, which in its main part had to determine if (in case of power loss at the unit) the electricity produced by turbine rotor *vybeg* would be sufficient to cover a gap in power supply to the reactor operational systems before the backup diesel generators kick in.

---

[2] The organization that governs the power distribution in the region.

For many hours one of the most powerful nuclear reactors on Earth was operating at half of its capacity with major safety system turned off.

The actual test was delayed until the early hours of April 26th.

At 11 pm on April 25th *Kievenergo* agreed to disconnect the unit 4 from the grid. The power decrease of unit 4 resumed.

New shift took over the unit control at midnight. Many of its workers would not live past the summer of 1986.

Meanwhile, L. Toptunov, Lead Engineer of Reactor Control of the doomed shift (basically, the main reactor operator), switched the operation of control rods from local to global option. This is one of the fundamental actions that was debated as a mistake in most sources analyzing the accident, including personal accounts, specialist analyses, and official documents. Transfer from local control option (insertion or withdrawal of control rods in the active zone using small groups) to global (when all rods are moved in concerted fashion) is a standard part of the reactor operation under certain conditions, and Toptunov did not violate the instructions. However, something went awry at this very moment, and the reactor power plummeted to only 30 MWt instead of the desired 700 MWt. Stability of the core was severely compromised: xenon poisoning was eroding the neutron flux, and together with a rapid graphite cooling, the situation called for urgent measures to bring the reactor back. Another option was to shut the reactor down and re-initiate the test when it is reactivated at later time. Dyatlov was nervous. He wanted to finish the test, which had dragged through two working shifts already. He ordered Toptunov to withdraw more control rods in order to "heat up" the core.

This point was pivotal in the chain of events that spiraled an initially unstable situation down to the catastrophe.

Should Toptunov resisted and acted with safety considerations in mind, the reactor would have been shut down. Yes, the test would be terminated. Yes, bringing the reactor from *iodine pit* back to operation would imply efforts that would not be welcomed because it would require a pause for a couple of days until xenon completes its decay. Yes, the operators would get punished, which — near the May 1st, Labor Day holiday — would have meant no bonuses. Yes, Dyatlov would give him hell.

But he would have saved the world from the worst nuclear disaster.

He followed the orders and began to withdraw more control rods.

It appears that the shift operators were either unaware of the high ORM (see Appendix A) of *RMBK* reactors, or were grossly underestimating this number. One way or the other, the total number of control rods that had remained in the core at 1:22 am on April 26th was either at the lowest amount or very close to that (by different sources, that number varies from as few as 6 to 8 to as many as 18), instead of 28–30 as recommended by safety specialists for *RBMK* reactors [1].

The time was 1:22:30. Power had reached 200 MWt, and Dyatlov decided to commence the test. At 01:23:04, the steam flow was cut off from the turbine (at that time, only one out of two turbines was connected to the unit due to the earlier request of *Kievenergo* to run the unit at 50%). Four cooling water pumps out of eight that were connected to the reactor began running down due to the lower power supply generated by rotor *vybeg*. The combination of diminished pressure in water feed, slower flow rate, and therefore increased temperature of the water, which was supplied to the bottom of the core, caused extensive void formation. Xenon poisoning, combined with rapid increase of PVC (see Appendix A), led to an outburst in the core reactivity. Akimov and Dyatlov were disturbed by the escalation of the power.

The SCRAM (AZ-5) button was pressed by Akimov at 1:23:40. Graphite tips of about 200 control rods, simultaneously entering the near-critical active zone, created a huge surge of reactivity due the positive SCRAM effect (see Appendix A). Instead of going through the channels all the way down, to the depth of about 7 m, the control rods were stalled at the depth of about 2–2.5 m. Panicked, Akimov disconnected the control rod servo motors, but the rods would not fall down on their own, as they should have. He realized that the channels were already warped, and the rods were jammed. At 1:23:43, *only three seconds later*, the reactor power reached 530 MWt and continued to rise exponentially. There was now much more steam than water in the core; fuel assemblies were ruptured, creating more overheated steam and resulting in even larger PVC. Loud rumbling noises were heard in the control room; they were coming from the reactor. Shock and disbelief tied up the hands of operating personnel. The reactor control panel showed that all pumps had failed, and there was no more water flow to the core. Yet the ECCS was disconnected, and its valves were chained and locked according to the test protocol to avoid inadvertent manual opening.

1:23:44, the reactor power surged 120 times over its full capacity.

The continued rupture of fuel channels and resulting overheating of the core increased the steam pressure in the reactor. Channel pipes began to rupture, causing generation of additional immense amounts of steam in the containment. Unbearable internal pressure blasted the reactor body apart; a loud explosion shook the whole plant. The massive (over a thousand ton) UBS lid flew up in the air, damaging the remaining technological infrastructure of the reactor. It dropped back down and was wedged inside, tilted at about 60 degrees. Ensuing depressurization and damage of water coolant circuit led to the generation of hydrogen as a result of the chemical reaction of zircalloy material with water. Oxygen of the air reached the piping hot graphite, which began to burn fiercely and ignited the hydrogen–oxygen mixture.

A violent second explosion followed the first one in three seconds.

Unit 4 RBMK-1000 was gone.

The explosions were so powerful that they crushed and spewed out the larger part of the reactor content — graphite, fuel, *TVELs*, channel material — both inside the reactor building and on the adjacent territory. Since the first explosion literally pulverized the upper building infrastructure and periphery of unit 4, nothing stopped the second explosion projectiles, which consisted of thousands of small and large pieces of the former reactor core, to fly out for large distances and to seed the territory of the ChNPP, from the rooftop of unit 3 to the turbine hall and to the grounds of the Industrial Zone around. The sheer force of the blast was so strong that many of the pieces went through the ceiling and the roof of a machine hall and landed inside the turbine hall. A huge fireball — almost spherical, about the size of the former reactor core, rose above the damaged reactor, and in a few moments disappeared in the large black plume, which gave many eyewitnesses and specialists a reason to believe that the second explosion was nothing short of a nuclear blast.

Looking from outside, the upper half of the unit 4 building was almost completely gone. A strong blue glow of emitted ionized air was coming out from the damaged reactor. Plumes of graphite-burned ash, smoke, steam and dust, all radioactive beyond comprehension, were rising in the dark night sky, hemmed by red reflections of multiple fires. The radioactive cloud reached the levels of about 1 km and was initially directed by prevailing winds toward North-West; heavier particulates and radionuclides were precipitating in the vicinity of the station, while the lighter ones brought up a radioactive fallout on most of Europe, as far as in Finland and Sweden.

Grey-white radioactive plumes were swiftly spreading inside the remaining hallways, technological rooms and facilities of unit 4 and even in unit 3. Radioactivity quickly reached and exceeded — in many places, by several orders of magnitude — lethal levels. Shift personnel in the reactor building, turbine hall machinists and electricians were scrambling to figure out the amount of damage, the actions necessary to contain the spread of fire and preventive measures to avoid further explosions. All this was done in almost full darkness (power was lost everywhere and working emergency lights were scarce), in suffocating, dreadful air filled with radioactive dust and steam, with water (not less radioactive) converging in many places in the buildings.

The first firefighting vehicles arrived at about 1:45 am — an incredible turnaround and reaction to the calls made by Akimov right after the explosions. Firemen had to fight multiple fires both on the rooftop of unit 3 (in several instances almost on the edge of the newly formed crater, where radiation was deadly enough for a person to receive a lethal dose in a few seconds), on the roof of turbine hall, inside the reactor building, on the ground ... Closer to 7 am, most fires — except a major one, the crater — were put out thanks to fearless persistent actions of over 190 firefighters from 37 fire brigades which were swiftly mobilized and arrived from Pripyat, Chernobyl and neighboring towns to help the ChNPP fire brigade. The majority of the casualties reported after the first days of the catastrophe were among the firefighters. They were dying one by one in Moscow hospital No. 6, the famed medical facility specializing in post-radiation treatment, where many Chernobyl-derived radiation sicknesses of plant personnel, firefighters, and liquidators were treated. Their bodies were emitting radiation so strong that they had to be buried in lead coffins in cement tombs deep underground.

Meanwhile, Dyatlov, in the firm belief that the unit 4 reactor was still in one piece and that the explosion was a blast of hydrogen accumulated in the emergency tank of large (over 100 cu. m) capacity, located under the central hall roof, ordered Akimov (who also believed in the undamaged reactor) to start pumping emergency feedwater into the reactor, greatly increasing the amount of radioactivity released by continuously produced steam and microexplosions in the crater of burning graphite that was a nuclear reactor an hour before. Aside from increasing amounts of volatile radionuclides, the highly radioactive water had now started accumulating in the cable tunnels in the lower levels of the crippled reactor building.

Dyatlov reported his beliefs as a reality to Director Bryukhanov and Dyatlov's supervisor, ChNPP Chief Engineer Fomin, and this false assumption badly tainted the first hours and days of the accident containment efforts.

The whole story of valiant efforts that were delivered, besides firefighters, by shift personnel in the reactor building, half of which was blasted into pieces and the rest was badly damaged, and in the turbine hall, who made sure that there would be no more hydrogen accumulation and respective explosions in the steam lines, truly deserves a separate book of praise. All of these guys were gravely burnt by radiation, receiving fatal doses and yet not leaving the site, continuing to work until they were unable to lift an arm, vomiting and losing conscience.

These efforts were in stark contrast to the firm and continuous orders that were given by Dyatlov and Akimov, and later by Fomin, who arrived at the control room of unit 4 to take lead in the containment efforts. Despite multiple reports from eyewitnesses about the reactor obliteration, many of which had received lethal doses just for several seconds of being in its locality or simply peeking through the slightly ajar door, despite the small projectiles — pieces of *TVEL*, graphite, and fuel — that densely covered a maze of the damaged reactor building remains and were instantly noticed by personnel sent by Dyatlov to assess the extent of the damage in the building, despite Akimov's own brave effort to go knee- or even waist-deep in highly radioactive water in the attempt to open the feedwater valves of spare coolant pumps, despite all that, the three of them, Akimov, Dyatlov, and Fomin, still believed that the *RBMK*-1000 of unit 4 was unharmed. Their fatal mistake was epitomized by the tragedy of A. Sitnikov, another Deputy of Chief Engineer Fomin. Sitnikov, who on the order of his boss went up to the roof and from several places did assess the damage of the central hall (much of it was already gone), looked down in the crater and saw with his own eyes that the reactor lid now rested inside the core and there unquestionably was a strong emission of radioactivity up in the open air. He then reported his observations to Fomin — and yet again, the Chief Engineer did not accept as true the words of his own Deputy. (Sitnikov received a dose so large that his body rejected multiple bone marrow transplants in hospital No.6 in Moscow; he died in severe pain among the first hospitalized there in April.)

So, Fomin and Dyatlov were sending people, the shift personnel, one after another to *manually* restore the feed of cooling water to the reactor that did not exist anymore.

Respective staff actions, no less heroic than any other work that was performed in the first few hours of the accident containment, but in this case ultimately worsening the situation and endangering the whole plant, resulted in the slow but continuous accumulation of radioactive water in the cable tunnels that were interconnecting four units of ChNPP and now endangering the safety of the other three reactors, because water was copiously pumped into unit 4 using at first the emergency supply of unit 4, and later emergency feedwater reserves of unit 3 and even 1 and 2. However, most of the pipes were already damaged or obliterated by the explosions, and water was mainly spreading along the damaged feedwater network, flooding and contaminating the plant facilities.

I would like to emphasize that under different circumstances, the persistence of ChNPP management in their hard work to restore the supply of coolant to unit 4 reactor would deserved nothing but acclamation. Cooling down the near-critical core is the main commandment in the nuclear reactor bible, because if there is no cooling, the ferocious energy of fission will eventually melt the fuel, steering the core melt-down. However, their assessment of the reactor status and their dogged fixation (lasting for hours if not days) on the false pretense that it was unharmed, have critically affected the containment and cleanup efforts for days to come.

Meanwhile, the news about the accident had reached the Kremlin in the early hours of April 26th. Deputy Minister of USSR Energy, A. Makukhin, who received an extremely sugarcoated report from ChNPP management (Bryukhanov and Fomin) about the accident, informed the Central Party Committee that there was an explosion in the upper part of the reactor hall at the unit 4 of ChNPP that led to partial structural damage of the central hall, reactor hall and some supporting facilities. The fire that accompanied the explosion was suc-cessfully put out at 3:30 am. The plant personnel were taking necessary measures to cool down the reactor. There were 34 people, firefighters and staff, hospitalized as a result of the explosion and subsequent fire. *There is no apparent need to evacuate the nearby town of Pripyat.*

(Pripyat was a town about 2 km from the station, built for the ChNPP workers and their families. There were close to 50,000 people who enjoyed living in this young (average age 26), bustling town. All of them were fully unaware of the tremendous radiation pouring out of the compromised reactor, continuing to go about their daily lives in the sunny Saturday morning, when the radiation levels on the streets were already reaching 5–10 R/h, and in some places substantially higher.)

The swift verdict came back with indisputable authority: continue pumping the water to cool the intact reactor.

Imagine that you try to cool down an industrial size blast furnace at full fury with a garden hose!

The Special Government Commission (SGC) was created, managed by Deputy Chairman of the Council of Ministers B. Shcherbina. The SGC included several high-ranking officials on the level of Deputy Ministers, leading nuclear scientists, representatives of organizations responsible for design and operation of *RMBK* reactors, health author-ities, military and police operatives. The Commission's task was to "liq-uidate the consequences of the accident efficiently and expeditiously". Shcherbina demanded a full and candid report on the current situation with the reactor, about the radiation conditions at the ChNPP overall, and in the nearby town of Pripyat. However, the true gravity of the situation had become apparent only at daylight of Saturday, April 26th. Only then, in the Saturday morning and afternoon, the full extent of the radioactive exposure was slowly getting assessed — unfortunately, not by everyone. Sadly, not by the residents of Pripyat and definitely not by the people in charge.

All these days, through the first week since the explosion, the crippled reactor continued to emit enormous amounts of radionuclides from the gutted crater. Plumes of smoke were drawing the attention of the Pripyat residents; some of them went fishing that dreadful night, some went to the improvised beach at the artificial lake next to the plant cooling pond. As happened later in Fukushima, the problem of accumulating the huge amounts of highly radioactive water at the site had to be resolved; but in Chernobyl there was no ocean nearby, and thus the heavily contaminated water was hosed down to the cooling pond, contaminating it over the roof. This extremely radioactive water was mainly drained from the basements and cable tunnels of the reactor building. My team had the misfortune to uncover some of those hoses during one of the clean-up jumps in August. The billow coming from the damaged reactor building and the partially collapsed central hall were clearly visible to Pripyat residents, and on Saturday morning the more adventurous youngsters went on bikes to the overpass bridge, coming across the local railroad, to check out the fire at the plant — the overpass was closer to the unit 4 than the town. At that time, the radiation on the bridge had reached the levels closer to 500 R/h because the initial, far more dangerous, radioactive cloud went through there. The teenagers acquired severe radiation sickness due to an exposure to high radiation fields on the bridge.

A grim shock came after the first helicopter flight around and over the damaged unit 4. Photographs made during the flight have since became iconic. It was then clear that the reactor was badly damaged and there was no more protection to contain the fierce radiation coming from the demolished core, which was complicated by burning graphite. The amount of graphite strewn around the crater was absolutely astounding.

By late Saturday, SGC was seriously considering the possibility of spontaneous nuclear reaction of the remaining fuel in reactor. Due to xenon decay, its sponge-like absorption of neutron flux should have been ending in about 2–3 days from the beginning of xenon poisoning, and sometime around Sunday to Monday night it should not be a factor anymore to contain the threat of the core meltdown. Later at night, the decision to evacuate Pripyat was made, and 1,200 buses were wheeled into town. All 48,000 inhabitants of the town were driven out of Pripyat on Sunday afternoon.

Forever.

They were told to take only valuables (money, passports) and medications for the trip — they were told that they would go back after three days. They left all their belongings behind. They left beloved pets. They left behind their happy, peaceful life, which would never be the same.

SGC started developing contingency plans, which were created literally on the go, because the damaged reactor continued emitting an enormous flow of radiation. The ChNPP management had finally arranged (after almost two full days since the accident!) systematic dosimetry control that established continuous radiation measurements around the unit 4, in the plant surroundings, and in places adjacent to the original radioactive cloud path.

The received data were plainly scary.

In order to block the reactor from sending enormous amount of radionuclide aerosols day after day, and simultaneously to prevent the possibility of spontaneous re-initiation of nuclear reaction, the bombardment of the destroyed reactor was performed, beginning as early as Sunday afternoon. It seemed to bring limited success as the radiation levels began to drop in the first days of May. However, analysis of the emission kinetics completed at a later time and especially radiochemical analysis of the ratio I-134 to I-131 that was performed in early May on a daily basis, showed that there was no change in the ratio of short-living iodine isotopes, which meant that the remaining core content was in sub-critical condition and there was no threat of chain reaction to occur. This, in turn, resulted in the slight but continuous drop of the radioactivity around the crippled reactor in the first week of May.

Nevertheless, in early May a new danger had emerged related to a threat of full or partial core meltdown (or whatever was left of it).

The thermal processes inside the runaway reactor core, where the temperature was unbelievably high (exceeding $2,000^\circ$ C) because of the absence of coolant, led to the melting of nuclear fuel, fission products, zircalloy, and other materials inside the active zone. That mass was capable of literally burning its way across anything beneath it (gravity drove it down), including steel and concrete. For the overheated core material of the runaway reactor, there was nothing that could stop it from eating out the safety shield underneath (Russian word *prozhig*[3] means the movement of melted content of the former core through the containment), flowing down, reaching the underground waters and dispersing incredible amounts of radionuclides over the whole aquifer system. Cooling the overheated core remnants had become a primary task in early May.

The first selected and approved option by SGC for preventive cooling was the creation of an interconnected pipe framework in the grounds under reactor building, where cool water will circulate, thus bringing the core temperature down. To serve this task, the pipe framework of 30x30 m size was designed and manufactured, and then 400 miners and subway diggers began relentless work on reaching the targeted underground space through the narrow tunnel. However, their gutsy efforts proved to be useless afterall, because in mid-May it became clear that full *prozhig* had no chance to occur (more on that a bit later).

Yet another threat seemed to be imminent.

The containment of *RBMK* was built above two vertically stacked so-called bubbler pools, sizable concrete-laid basins for condensing and containing the excess of overheated steam during emergency reactor cooling. Since the initial calculations assumed that the majority of the core mass was still inside the damaged reactor, the probability of *prozhig* content that would pave its way to the bubbler pools was very high. They presumably contained unknown amounts of water, collected there during the emergency cooling at the time of accident,

[3] Burn-through.

plus potentially some feedwater that was continuously pumped in large excess to cool the "intact" reactor after the accident. Possible contact of the superhot *prozhig* material with water meant another explosion, "steam explosion", with much graver result (some credible sources say that the SGC's decision on creating the 30 km Exclusion Zone was primarily driven by the possibility of steam explosion [3], and that once this threat was eliminated, the necessity of such wide Zone was doubtful). It was necessary to check on the presence of water in both bubbler pools, and if there was water, to drain it elsewhere from underneath the reactor core. The task was successfully completed by the team of volunteers on May 3rd and 4th, who first determined that most of the water was located in the lower bubbler pool, which was then drained using firefighting equipment. All volunteers received exceedingly high doses, but were quickly treated with relative success for radiation sickness (this was definitely an act of calculated heroism, because these guys knew what they would face under the damaged reactor, and yet they took upon the noble duty).

To provide further efficient cooling of the crippled reactor from beneath, SGC approved the assembling of a liquid-nitrogen-fed pipe system in the bubbler pools, which was partially completed; however, it was the second half of May when a new phenomenon was discovered. The liquefied core mass indeed started chewing the bottom part of the reactor vessel, flowing down in three streams to the technological rooms at the level of the upper bubbler pool and even reaching the lower pool (by some calculations, about 14 tons of radioactive fuel escaped the reactor [3]). But something very strange happened — while melting the construction materials, the overheated lava reacted with their content (silicon dioxide, magnesium oxide and others) and created a substance that was christened as "lava-like fuel containing material" (LFCM). This chemical reaction was apparently endothermic enough to efficiently cool down the "corium" (another word, often used in modern Chernobyl slang for describing LFCM), without the need to involve the external cooling by liquid nitrogen, nor by water. The most studied *prozhig* material that exited to the bubbler pool area, eventually solidified into a shiny black glassy mass that resembled an elephant foot (it was discovered much later), and was called this way ever since. Initial assessment of the radiation level of the "Elephant Foot" was between 8,000 to 10,000 R/h. Even now, after decades, it still emits extremely dangerous radiation.

The last severe eruption of the damaged reactor happened in the evening of May 9th, when dosimeters registered a sudden jump of radioactivity as far away from ChNPP as in Ivankov. As specialists reasoned, that was apparently a result of graphite burnout in the lower part; then, the upper portion of the crippled reactor — rubble, collapsed structural elements, solids from the bombardment — crashed into the newly formed void and hit the corium. Eyewitnesses reported that a surge of blue rays, sparks, and plumes of black and white smoke billowed from the crater, covering land that was already suffering from nuclear devastation around ChNPP with yet another layer of radioactive ash.

Smaller, less serious microeruptions ("puffs", in liquidator slang) continued to happen far longer, until August–September of 1986, when the containment (the sarcophagus, in liquidator slang) covered unit 4 completely. They did not change the radiation profile much on some distance from the plant, but were giving a continuous headache to liquidators, who were struggling to decontaminate once and for all the industrial site around the main reactor building, since each of these puffs dispersed another round of radionuclides on just decontaminated surfaces, which had to be cleaned up over and over.

In early May, three zones around ChNPP were created, first one (Special Zone) around the plant itself, second (10 km Exclusion Zone) encircling the areas from which inhabitants were evacuated during May 2nd–3rd, and third (30 km Exclusion Zone) was mainly vacated on May 4th–7th. During the first few hours and days of the aftermath, the government and SGC mobilized thousands of policemen to patrol Pripyat and ChNPP, to guard the plant and to cordon the Exclusion Zones. Amazingly, frequent looting that later became a staple for abandoned villages and towns in the 30 km Zone, was not as prominent in the early days of *the war*, as locals called the accident and its aftermath.

Massive involvement of military forces, which was ordered by the government in the first weeks of the aftermath, was initially erratic and poorly organized. Stunning episodes of bravery were interlaced with ignorance, nothing short of stupidity, when soldiers in field uniform and without even marginal protection were ordered to pick up pieces of scattered graphite with bare hands and dump them in the collection containers around the demolished reactor. The worst part of this incredibly bold and yet inane engagement was that those troops were *regular Soviet Army* soldiers, all in their 20s at best if not younger; they were burned and overexposed to high doses in hundreds, and very quickly. The long-term effects of high energy radiation fields on the ability of these young men to have healthy children were not known to doctors in full details, but for sure the prognosis was not overwhelmingly cheerful. So, those youngsters had to go home. The Ministry of Defense had to make a critical decision of calling in the Army Reserves. The threshold was set at 30 years and older — apparently, this was considered a *"had your kids already"* age.

A few weeks later, I arrived at the 25th Brigade of Chemical Defense and began my work as a liquidator.

Factors That Triggered the Accident and Affected Its Aftermath

I wanted to mention several reasons that I personally think were pivotal for the catastrophe to happen (in fact, to make it almost unavoidable), and also affected the magnitude and the efficiency of cleanup efforts; they could be defined as human, sociopolitical and technological factors. I list them in no particular order, highlighting their overall importance for the accident and its aftermath, where applicable.

1. The veto on publicly disclosed information about the threats of poorly managed nuclear energy, and simultaneously the trumpeting of "safe" and "rock-steady" nuclear power technology.

For many years, nuclear energy and all related spheres of industrial and military importance in the USSR were under the tight control of an establishment with almost unlimited capabilities, yet with an obscure name — *Minsredmash, Ministry of Middle Machinery,* in the very beginning headed by Lavrentyi Beria, the notorious Chief of Stalin's KGB. The KGB-style *modus operandi* of this monstrous organization has not only suppressed the rightful concerns of the balance of pros and cons in nuclear energy use, but also masterfully, and for decades, kept fostering an image of the safe and foolproof "*peaceful atom*" in the minds of Soviets.

2. The gag on information about accidents that had happened by the dozens in the history of the USSR nuclear energy use.

Even for nuclear power plant personnel, who could have definitely learned a lot from the previous accidents and their analyses, this information was out of reach.

3. The industry-wide approach that inevitably corroded the actions of the operating personnel during the ill-fated test of unit 4 — it was not written anywhere but was fully adopted throughout the whole energy community: *power before nuclear safety*.

Released in 1991, the root cause report by the *USSR State Committee on the Supervision of Safety in Industry and Nuclear Power* (SCSS-INP) contained a direct affirmation of this approach: "*the operating algorithm of the emergency protection system ... was considered ... in terms of the efficiency of the plant's operation in the power supply system, rather than in terms of its ability to ensure nuclear safety"* [2]. This indeed was applicable to the whole energy sector of the USSR nuclear industry.

4. The patchy professionalism, experience and knowledge at all levels of the nuclear energy industry, and at ChNPP in particular.

Because the nuclear power facilities were regarded first and foremost as the *electric power plants*, the vacant engineering and management positions requiring a thorough expertise and multi-year experience were often filled by specialists from the neighboring energy sectors (for example, by personnel from thermo- or hydro-electric power plants), or even by the Party "functionaries" — a trait that was typical for the Soviet industrial revolution beginning in the early 20th century, and that was widely accepted all the way through the end of the USSR. The main snag of such "reinforcements" was that at best they were familiar with the power production part of their duties, but not with the nuclear energy principles and nuclear safety, or even worse, had no knowledge of the science and technology involved, but were great "follow-through" managers with an unblemished Party reputation, incapable of saying "no" to a directive coming from the top, no matter how questionable or plainly wrong they were.

5. The inadequate level of technological documentation, standard operating procedures, and protocols related to the functioning of nuclear reactors under duress; the lack of appropriate contingency plans for personnel involved in handling an unstable reactor, particularly of *RBMK* type.

As I mentioned a few times here, the complacency of some ChNPP plant personnel, often bordering on ignorance, was driven by two factors: a firm belief in the wide fool-proof operational margins of *RBMK-1000* and in the *"electricity first"* tactic.

For example, the positive scram effect was for the first time observed and studied at *RBMK* unit 1 at Ignalina NPP *and also at ChNPP —at the same unit 4* (!) in 1983. Respective reports were submitted to the upper management, including recommendations to alleviate it by altering the design of the control rods (removing the water gap in the bottom of the core upon fully retracted control rods), but these reports had never made it to the country's nuclear community, and no actions took place. Of course, there was no evidence that ChNPP unit 4 operators were unaware of these findings, but I am sure that if they knew the full extent of this effect, they would have taken a different approach to regain control over the runaway reactor.

Respectively, there were no relevant, comprehensive, and cohesive contingency plans of actions of nuclear reactor operating personnel during refractory situations. As an example, the INSAG-7 Report stated that *"the regulations 'General Safety Provisions GSP-73' and 'Nuclear Safety Rules NSR-04-74' came into force more than ten years before the accident, during which time Chernobyl (unit) 4 was designed, constructed and put into operation. During all that time, neither the chief design engineer, nor the general designer, nor the scientific manager took effective measures to bring the design of the RBMK-1000 reactor in line with the safety standards and regulations"*. [2]

Why did this happen? — The country needed electric energy.

6. The authoritarian methods of assessing the situation, decision-making, and resolving the problems before, during and after the accident on all levels of management and the government/Party establishment.

I would like to talk about this factor in detail. This is perhaps the most typical that can be said about the accident and its aftermath in general, and I must admit that it was perceived as the right way of efficiently taking care of *anything* in the USSR, beginning with a decision to enter into war to a choice of a kid's face on a candy wrapper. Thus, when the initial shock of realizing that something of planetary scale (and not in a good sense) had happened, the same old millstones started moving with a screech, and a gigantic colossus of authoritarianism was awakened yet again. It showed up at every level, before, during, and after the catastrophe, beginning from the decisions to alter the test protocol for unit 4, followed by the persistency with which the initial measures to normalize the situation were based on the deep-rooted belief of the Deputy Chief Engineer of ChNPP N. Dyatlov that the reactor was intact and it was imperative to pump the coolant (water) into it as a desperate attempt to contain the runaway situation (while in reality the two consecutive explosions had already demolished the reactor).

The same belief was the reason to send dozens of firefighters to put out multiple fires on the rooftop of unit 3, adjacent to the exploded unit 4, and even inside the unit 4 building (orders that essentially meant a death sentence). The same belief was disseminated across the giant pyramid of local, Republican, and State establishment; it initially gave the government and Party bosses a reason to lie to the country and to the world that the damage was minor and the situation was under control.

One more epic example of the authoritative folly was the decision to bombard the open crater of the damaged unit 4 with various "ancillary" materials from helicopters; it was suggested that such "clogging" would curb the radiation spread by filtering the radionuclide emanation, decrease the fission, and blanket the burning graphite. Lead, sand, boron and dolomite were dumped in tons by multiple helicopter raids over the open reactor crater. It failed on many levels, because the "precision drop", which was initially attempted with choppers hanging over the crater, quickly decimated the available helicopter crews: a powerful, ferocious discharge of radiation from the reactor straight up in the air for hundreds of meters was deadly (the initial drop was planned at the 100–110 m height, but the radiation there already exceeded 500 R/h, and a few sand bags that were dropped in the fiercely burning crater instantly elevated it to over 800 R/h), and yet the pilots had to stick their heads out of the windows, eye-balling the drop off point. Pilots were stressed by exceedingly high radiation and close proximity of the deaerator stack and construction cranes — one helicopter collided with the crane and fell, killing the crew of four. Aerial photographs of unit 4 taken around mid-May 1986 clearly show how "precise" was this approach: a large part of the bombardment missed the target, crushing the roof of the turbine hall, the roof of unit 3, breaking parts of the remaining wall and floor structures of unit 4. Some sources say that the weight of the dropped materials had pushed the UBS lid further down in the crater and tilted it even more, which increased the release of radionuclides in the atmosphere. Most painfully (because many pilots and crew members were severely overexposed in a matter of the first few hours and had succumbed to radiation disease in the following months), these actions in some way created the opposite result. To imagine what effect a heavy, several hundred kilograms, lead roll would produce when dropped directly into a blaze of burning graphite, think of a big rock that is thrown into a bonfire: it brings up a lot of plumes, sparks, and essentially for a short moment accelerates the burning.

The emblematic use of *"biorobots"* is yet another example; these were the liquidators charged with an epic task to clean up the debris and to remove the roof cover of unit 3, which was severely contaminated by the unit 4 explosion products. They were sent in massive numbers (up to several hundred "nuclear jumpers" per day) to work in the radiation fields strong enough to kill a human in a matter of minutes. This is a typical case of the aftermath management in *Kremlin style*. Because of the extremely high levels of radiation on the roof of unit 3, where a substantial part of the exploded reactor content was scattered, initial attempts to remove the radioactive rubble were focused on the use of remote-controlled robotic equipment. However, that option failed quickly, as the telemetry was jammed by high radiation fields, and most of the robots stopped working. Additionally, moving those robots, which were able to keep up with high radiation, from one place to another on the roof of difficult terrain was a logistic nightmare. Then a decision to clean up the roof *manually* was made, and hundreds after hundreds of liquidators went through the dreadful exercise of cleaning the surface of unit 3 rooftop — literally by hand. This is perhaps the most tormenting chapter in the liquidator movement; it coined the meaning of the term *"biorobots"*, a grim nickname that was used among liquidators in 1986. Multiple places on the roof of unit 3, including its top level and all five platforms of deaerator stack, were densely covered with the reactor "stuffing", including in some cases almost intact fuel element assemblies. These remnants were cleaned up by the hands of the *"biorobots"*. The radiation levels near those objects were reaching thousands of R/h.

7. The lack of infrastructure, measures, and personnel prepared and trained to deal with emergency situations specifically related to the nuclear safety.

The ChNPP — just as the rest of the Soviet nuclear power industry — had no emergency tactical teams equipped and trained to provide containment and search-and-rescue operations in the event of radioactivity release; there was no training for plant personnel in relation to staff actions in nuclear crisis situations. The doomed shift on the night of the accident had no protective gear (even rudimentary), no abundant portable dosimetry instrumentation, and no way to interact with each other while frantically running through the maze of the semi-collapsed reactor building, filled with steam, dust, ash, and water — all highly radioactive — in desperate attempts to assess the damage, to search for their teammates, and to manually initiate the water feed for reactor cooling (again, because of the fatal belief of the Deputy Chief Engineer Dyatlov that the reactor was intact). All components of automation and communication networks were already wrecked after the first explosion, and the power went out; in addition, the operators had no way of knowing the scale of damage and radiation levels because after the explosions they lost the connection with all stationary detectors and probes in and around the reactor. The shift dosimetrist had only two portable dosimeters, one with a measurement limit at 1,000 R/h, another at *1,000 mR/h* (1 R/h); the first one malfunctioned quickly, and thus the only dosimeter left for assessing the actual radiation levels was useless because pretty much *everywhere* in the building the levels were much, much higher than 1 R/h. There were no radiation first aid kits containing *antirads*, specialty medications for emergency treatments of personnel affected by radiation exposure (potassium iodide for one), anywhere throughout the premises of reactor building.

The firefighting personnel (the first response was fronted by over 60 men), who were called upon right after the first telephone call from the operation room requesting to put down the fires, were fully unprepared and, just like the crew of the unit 4 shift, had no special protective gear to operate under nuclear emergency conditions. Meanwhile, the fires, totaling about 30 sources of various magnitudes, were spread over a large area around the demolished unit 4, because the lava-hot content of the exploded reactor was dispersed on the roof of adjacent unit 3, on various surrounding structures and on the grounds around the burning unit 4 reactor. They heroically fought the fire, walking over the scattered pieces of radioactive fuel, graphite and control rods that were emitting deadly radiation, and some of the firefighters were even picking up the broken pieces, trying to figure out what these things were and how did they end up there ...

8. Multiple alterations of design and violations of codes and regulations during construction of the ChNPP; the corner-cutting attitude and "cost-effective" measures.

A lot has been said and written about these issues in the analysis of the accident. Perhaps, one of the biggest mistakes was the choice of location. There were lengthy debates in the government and Party leadership — is the Chernobyl region a good area for building a nuclear power plant of such capacity? The reasonable and appropriate concerns of some nuclear energy experts, related to the safety and potential hazards of building a nuclear facility too close to Kiev, one of the largest cities in the USSR, were drowned in the greed of getting another powerful and cheap source of electricity in the European part of the country. Meanwhile, the Polesye region and Pripyat River basin were definitely a bad choice for this, not only geographically; the Pripyat aquifer consists of fine to medium grain quartz sands (depth of 18–22 m), easily permeable for groundwaters and positioned on a bed of dense, practically impenetrable, Quaternary deposits [4]. That presented a threat of creating a pathway for radionuclide migration (in case of trouble) to the Dnieper River aquifer, a major water course for Ukraine, with over 32 million people residing in Dnieper basin. Kiev, the capital of Ukraine, with a population of over 2 million people, is a mere 100 km away and situated on Dnieper River below the Pripyat merge.

"Improvements" in the reactor building design that were done to reinforce the structural integrity and alleviate the potential leak of radionuclides into groundwaters, did not look impressive on the background of other multiple violations in the insulation of aquifer from potential contamination; for example, the cooling pond was built with a levee from highly permeable local sands and did not include a low-permeability liner. This design resulted in extensive seepage of highly contaminated water through the levee in the years following the accident, with yearly seepage losses from the pond estimated at one half to two thirds of the total reservoir volume [4].

Yet another example of cost-cutting measures was the infamous rooftop of unit 3. By fire safety standards, the top cover of the roof must be made from fire-resistant materials, but ChNPP reactor unit rooftops in many places were covered by ignitable, easily melting bitumen, and this created an incredibly challenging problem of radioactive rubble removal. After the explosion, flying pieces of red-hot graphite, fuel, rods, broken building rubble, and demolished technological odds and ends were blown out of the reactor and dispersed on the rooftop. The heat from the damaged reactor and the burning graphite started multiple local fires there, which were heroically put down by the firefighters, but the bitumen softened and melted in many places, to the point where heavy pieces of whoppingly radioactive materials were sucked in the melting tar like in a swamp. Once the fires were taken care of, the temperature of the cover decreased, and the bitumen hardened again, this time coating the drowned pieces of radioactive debris inside the solidified cover. In essence, the removal of radioactive debris from the unit 3 rooftop in many places seemed like a deadly version of "hide and seek", where the liquidators had to identify the hidden "treasures" in the rooftop cover, mark them, and then physically remove the resinous bitumen with highly radioactive remnants of reactor core wedged in it, employing rudimentary equipment (wheelbarrows, shovels, axes) and often using bare hands.

9. The decision to restore the functioning of the crippled ChNPP, instead of entombing the whole plant once and for all.

I have read hundreds of various sources related to the Chernobyl catastrophe, and not a single one of them had an answer to the question of why on earth the Soviet government chose to continue using the heavily contaminated ChNPP after the accident.

Let me answer this query.

The actual motif is as clear as daylight, yet none of the numerous books, articles, proceedings of the conferences, symposia, meetings and so on about the accident had looked at this fundamental question, considering the scale of clean-up efforts and related cost.

Think about this concept for a moment.

I always admire the bravery of people who live in the low-lying peripheries of big rivers or near active volcanoes. Every now and then there is trouble — flooding, eruption, earthquake — and yet they stubbornly come back to the same places that gave them enormous problems, often taking countless lives. However, they return, because there is more to it than just a material loss or even loss of lives.

But there was no reason for my former country to make a similar comeback, to re-initiate the operation of the Chernobyl plant — except one.

The country bosses desperately wanted to prove to the world that all was fixed, and fixed efficiently, business as usual, and USSR was as mighty as always.

What I am getting at is that the ChNPP was eventually shut down. It means that all those tremendous efforts, hundreds of lives, enormous expenditures were worth nothing because it was closed down anyway! I am sure that the country leaders would have applied the same tenacity in fixing it, even if the plant was scheduled to work *for just one day* after the cleanup completion.

We needed to convince the world that all was good in the empire.

Should the government decide to shut the plant down from the very beginning, the whole Chernobyl saga would have developed in a totally different path, at least for us, liquidators. Think how many tasks would have been completed easier, faster and more efficiently, without risking liquidators' lives. How many repetitive, dangerous everyday jobs would have become obsolete... This thought is so disturbing and depressing, that I let it rest without further comment.

Bibliography

(1) Gr. Medvedev (1991), *The Truth About Chernobyl* (New York: I. B. Tauris).

(2) INSAG-7 (1992), *The Chernobyl Accident: Updating INSAG 1; A Report by the International Nuclear Safety Group* (Vienna: *Int. Atom. En. Agency*).

(3) R.V. Arutyunian (2018), *Chernobyl — Fukushima: Travel Notes of Liquidator* (Moscow). (in Russian).

(4) D.A. Bugai (1997), The cooling pond of the Chernobyl nuclear power plant: A groundwater remediation case history, *Water Resources Res.*, **33** (4), 677–688.

Index

Illustrations

Picture 1: *One of the first aerial views of the damaged reactor. Steam and smoke are coming out of the crater. The white multistorey cube at center left is ABK-2. The chimney to its left is the deaerator stack of Units 1 and 2; they can be seen further behind ABK-2 (black buildings with two white stripes).*

Picture 2: *Satellite view of the demolished Unit 4. The damage to the walls of Unit 4 and to the structural elements of turbine hall as a result of helicopter bombardment with ancillary materials is visible.*

Picture 3: *Aerial view of the damaged reactor (mid-May 1986).*

Picture 4: *Preparation for the jump to "The Roof".*

Picture 5: *"Mogilnik" of military and civilian equipment contaminated during the clean-up work at ChNPP.*

Picture 6: *Author in 1985, about a year before the Chernobyl stint (Dnepropetrovsk, Ukraine).*

Picture 7: Author in August 1986 (25th Brigade of Chemical Defense).

Picture 8: *Author's pass "CHERNOBYL", 1986 (valid until 09.30.86); the photograph shows signs of extensive damage of the silver halide layer due to the radiation.*

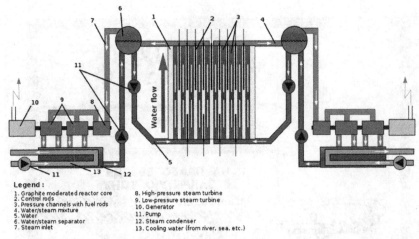

Legend :

1. Graphite moderated reactor core
2. Control rods
3. Pressure channels with fuel rods
4. Water/steam mixture
5. Water
6. Water/steam separator
7. Steam inlet

8. High-pressure steam turbine
9. Low-pressure steam turbine
10. Generator
11. Pump
12. Steam condenser
13. Cooling water (from river, sea, etc.)

Picture 9: *Schematic of RBMK-1000 reactor.*

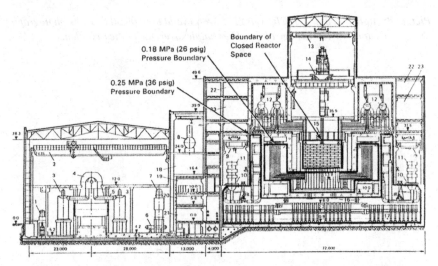

1 - First-stage condensate pump; 2 - 125/20-t overhead travelling crane; 3 - Separator steam superheater; 4 - K 500-05/3000 steam turbine; 5 - Condenser; 6 - Additional cooler; 7 - Low-pressure heater; 8 - Deaerator; 9 - 50/10-t overhead travelling crane; 10 - Main circulating pump; 11 - Electric motor of main circulating pump; 12 - Drum separator; 13 - 50/10-t remotely controlled overhead travelling crane; 14 - Refueling mechanism; 15 - RBMK-1000 reactor; 16 - Accident containment vavles; 17 - Bubbler pond; 18 - Pipe aisle; 19 - Modular control board; 20 - Location beneath control board room; 21 - House switchgear locations; 22 - Exhaust ventilation plant locations; 23 - Plenum ventilation plant locations

Picture 10: *Detailed cutaway of RBMK-1000 reactor building (right) and turbine hall (left), depicting main technological elements.*

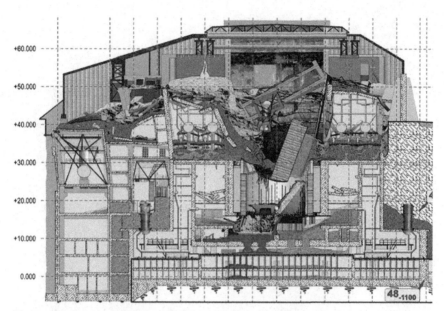

Picture 11: *Cutaway of destroyed ChNPP RBMK reactor, entombed in the sarcophagus: tilted UBS lid (sage-green) and prozhig (corium) material (red) are shown.*

Printed in the United States
By Bookmasters